CORSICANA
and the
ENNIS SUB

SP/SSW IN TEXAS

ROBERT P. OLMSTED

CORSICANA AND THE ENNIS SUB

Printed and bound in U.S.A.
Photography/Robert P. Olmsted
Halftones/Jim Walter Graphic Arts—Beloit, Wisconsin
Typesetting/Guetschow Typesetters—South Beloit, Illinois
Printing/Terry Printing Inc.—Janesville, Wisconsin
Binding/Zonne Bookbinders—Chicago, Illinois

Maps/Mike Schafer

Library of Congress Catalog Card Number: 83-62558

ISBN 0-934228-12-4

First Printing—November 1983

Acknowledgements: In my years of rail photography, which began seriously about 1948, I seldom have encountered an entire group of crews which are more friendly, helpful and informative than those on the SP/SSW in the Corsicana/Ennis area. My special thanks to all of them who aided me in this project. Thanks also to Joe Strapac, author of an exceptional series of "Reviews" of Southern Pacific motive power. The data herein was verified against Joe's thorough work; however, any errors are the responsibility of the author in preparing this volume. Additional thanks of course to Joe McMillan for getting this book into print and on its way to your easy chair.

Title Page . . . Consecutively-numbered 7674-75-76-77 work north through Corsicana, Texas on the Ennis Subdivision in December 1982. Eyeing the EMD GP40-2's, new in 1980, is GP9E 3734, a 1972 graduate of the Sacramento Locomotive Works GP9 upgrading program.

McMILLAN PUBLICATIONS, Inc.
3208 Halsey Drive
Woodridge, Illinois 60517

CORSICANA & THE ENNIS SUB

CONTENTS PAGE

SP trackage from Hearne through Groesbeck, Corsicana and Ennis to Dallas, and also the Forth Worth Branch, is all part of the Ennis Subdivision of the San Antonio Division in the 1980's.

SSW trackage into Corsicana from the east, and also the Waco Branch, is part of the Corsicana Subdivision of the Pine Bluff Division in the 1980's.

KERENS-CORSICANA

Left and above . . . On Thanksgiving Day 1981, Southern Pacific SD45T-2 9328 is in the hole at Kerens, Texas, the first long siding east of Corsicana on the Cotton Belt. Approaching on the main is a fast-moving, Corsicana-bound *Blue Streak Merchandise (BSMFF)*, led by the 8538, a 1979 Electro-Motive SD40T-2. Hordes of EMD's 3600 horsepower, twenty-cylinder SD45's and SD45T-2's, over 600 of them, joined the SP and SSW roster between 1966 and 1975. Subsequently the roads turned to the 3000 horsepower, sixteen-cylinder SD40T-2, such as the 8538, and GP40-2's.

Above and lower right . . . On another November 1981 morning, 8506 and GE 7871 crawl down the siding at Kerens with a lengthy merchandiser, for a meet with three southbounds.

Upper right . . . Four consecutively-numbered SP GP40-2's 7658-7659-7660-7661, roar through Powell, Texas, with a southbound *BSM* in January 1983. Cattle, cotton, oil and pecans are, or have been, the important commodities of this part of Texas. The peaks of importance for Longhorns and oil have passed of course. At Powell, a few miles east of Corsicana, the huge Powell Oil Field was discovered in 1905. A big oil boom began here, following a gusher which blew in January 1923. Over 40 million barrels of oil were pumped from a forrest of derricks in 1924. Now, in the 1980's, the derricks are gone and only a relative trickle of oil still comes from a handful of small rigs. There is no longer even a rail siding in Powell.

Left . . . General Electric 7871 at Kerens, Texas. Sixty B30-7's, 7824-7883, were built for the SP in 1979.

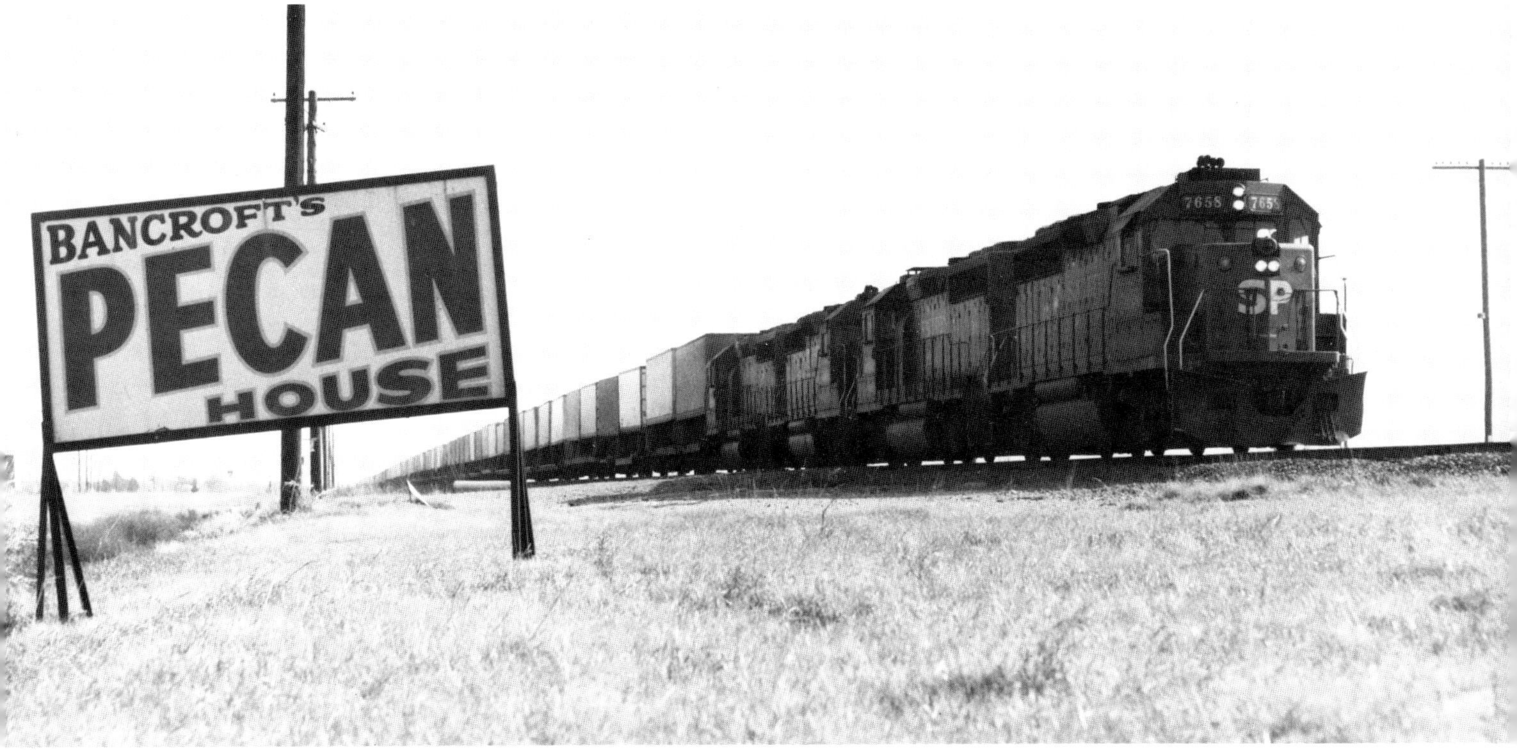
BANCROFT'S
PECAN
HOUSE
7658
SP

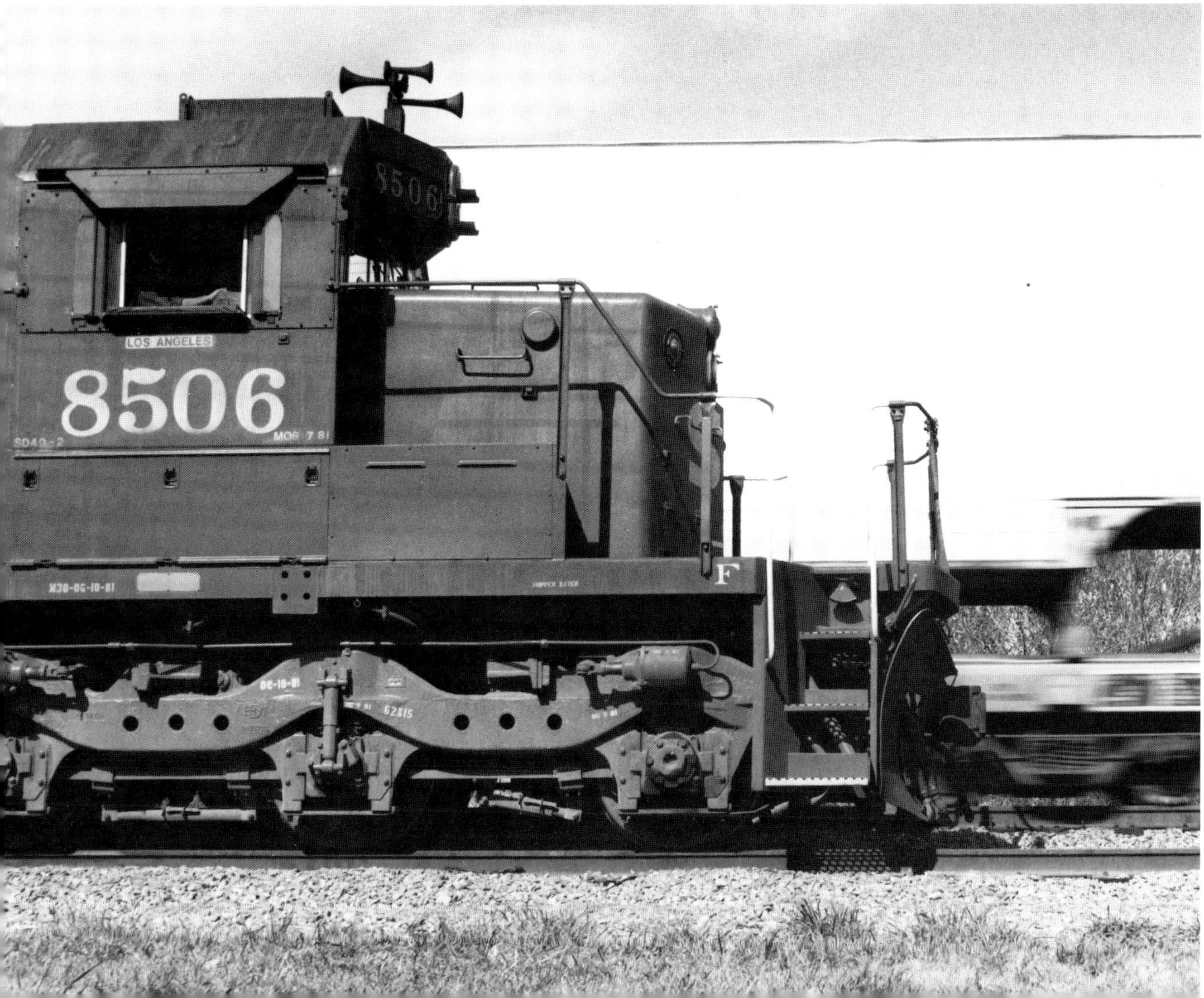
LOS ANGELES
8506
F

CORSICANA

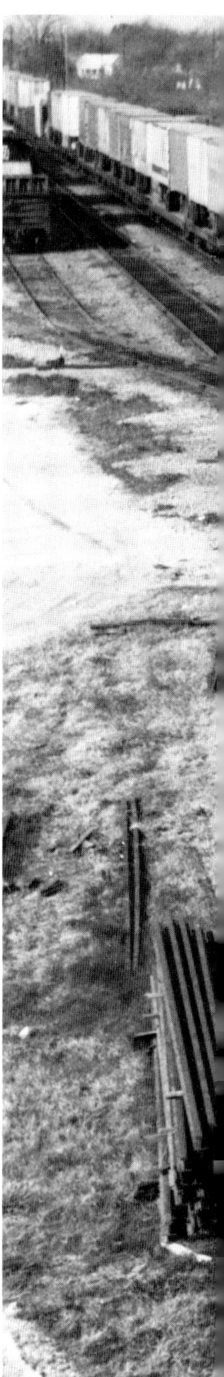

Above . . . A six-unit mix of SSW and SP diesels pulls past the small Hill Yard on the east outskirts of Corsicana in November 1981. Following a token purchase of eight GP40's (7600-7607) for the Cotton Belt in 1966, the SP stayed away from the four-axle locomotive until the late 1970's. Then seventy of EMD's GP40-2's, numbered 7608-7677, and including SSW 7638 leading the Tyler and Pine Bluff-bound freight, were delivered between 1978 and 1980. Additionally twenty GP40-2's 7940-7959 and over one hundred four-axle General Electric B30-7's were added to the roster between 1978 and 1980. With the exception of the 7900's, which are intended to operate with road slugs, these units all were ticketed for duty on the eastern lines (east of New Mexico).

Left . . . Cotton Belt 7785 and a trio of EMD's hustle *BSM* past a northbound waiting to head out the single-track main in December 1982.

7781 7781
SSW
7781

Above and left... SD40T-2 8307 works its way around the curve at the end of the long siding extending east out of Hill Yard. Shortly, red-nosed 7781 appears, leading one of the hottest trains on the Cotton Belt, the southbound *BSM. Blue Streak* service began on the St. Louis Southwestern in 1931. The 8300-series are extended-nose versions of the SD40T-2 housing equipment for Locotrol, radio control of remote, unmanned units in the train. GE's 7774-7799 are 1980 versions of the 3000 horsepower B30-7, assigned to the SSW. This particular January 11, 1982 was the coldest day of the Texas winter of '81-'82 in this area, with a morning low of 5 above and an afternoon maximum in the twenties.

Below... At the east end of Hill Yard, SD40T-2 8567 is on the point of a southbound awaiting a new crew. Sweeping by on the outside is the 7657 which will pick up a fresh crew on the other side of town. SSW 7657 is the highest-numbered of the thirty GP40-2's (7628-7657) in class EF430C-2, delivered to the Cotton Belt in 1979.

Below . . . Upgraded SD40E 7316 commands a four-unit EMD lashup approaching Hill Yard in Corsicana. In 1980 and 1981, all eighty-six remaining members of the group of SD40's built in 1966 and 1968 were upgraded at the SP's Sacramento Locomotive Works. SD40E 7316 was originally the 8445.

Upper right . . . Of the 340 GP9's built between 1954 and 1959 for the SP/SSW and the Texas & New Orleans, only twenty came from the builder with cut-down noses. These units, built in August/September 1959, were the only low-nosed GP9's produced by EMD. In February 1982, 3712 (formerly the 5876) is working Hill Yard.

Lower right . . . Cowboy hats abound as crews exchange greetings in Corsicana's Hill Yard. After the meet, the southbound, led by GE 7859, will work the yard. Two-year old 7859 was one of the sixty B30-7's (7824-7883) built in 1979, following 7800-7823 of 1978.

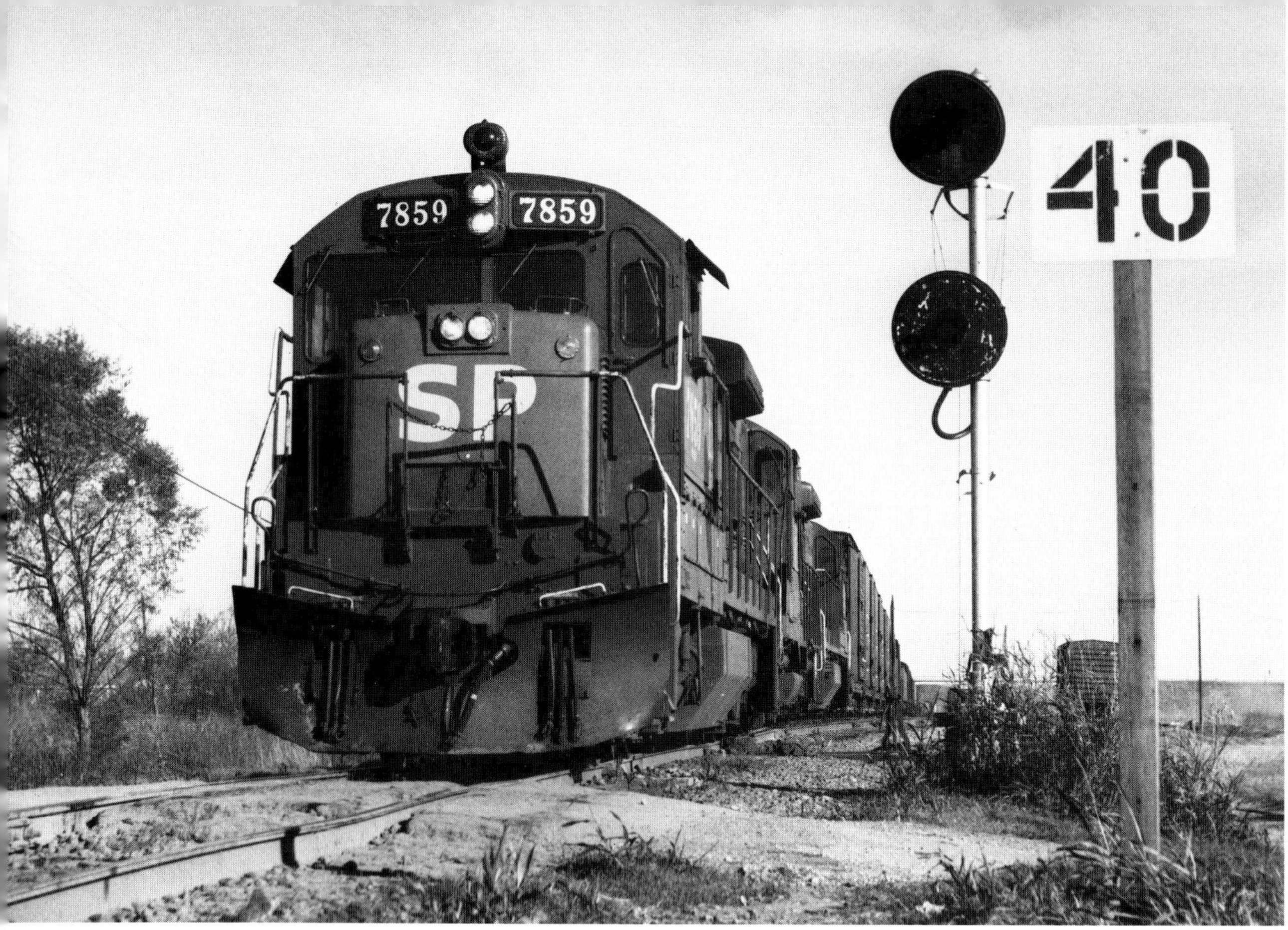
7859
7859
SP
40

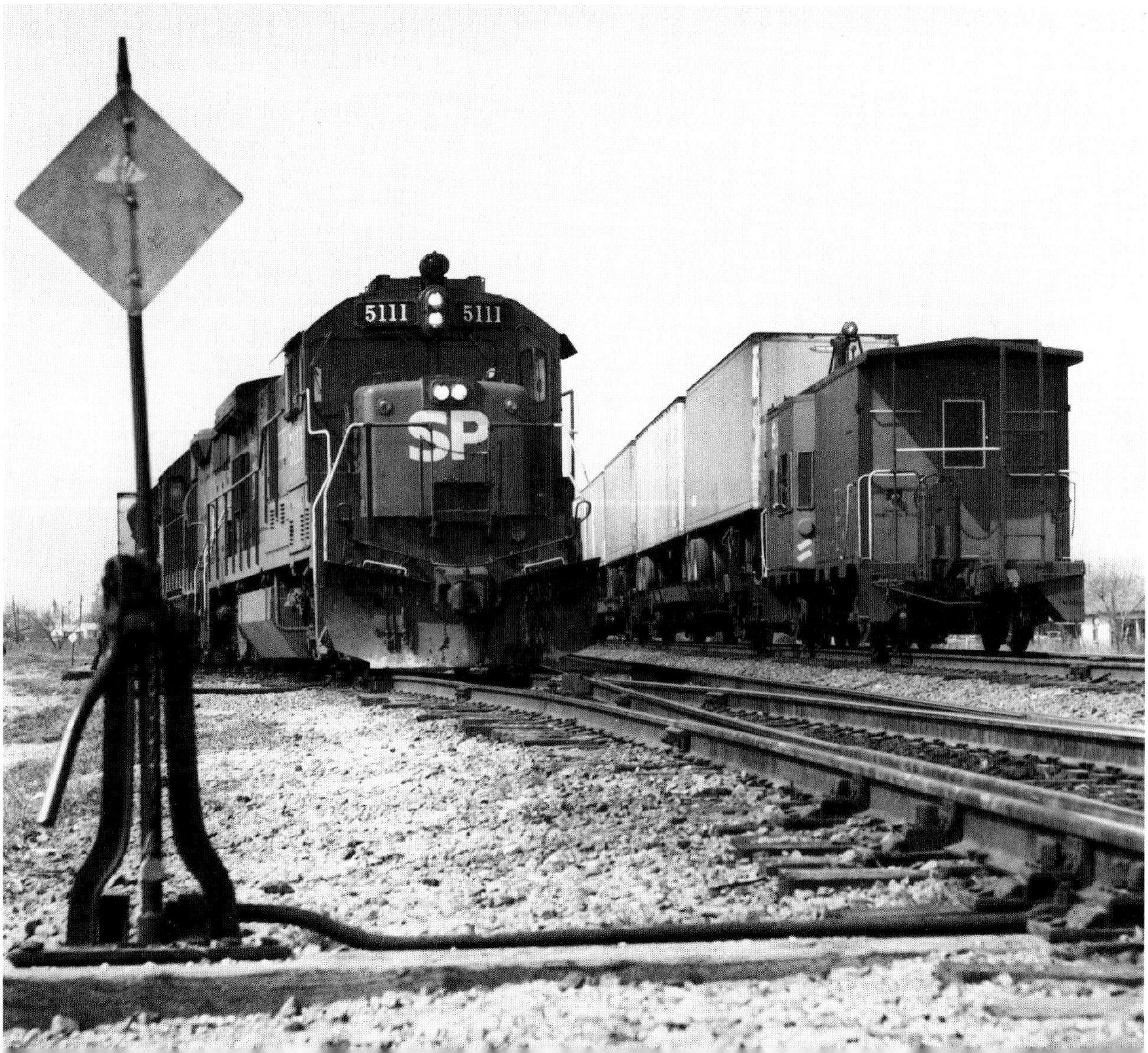
5111
5111
SP

Upper left . . . 7859 and its companion GE have set out cars for Waco; then picked up Geep 3381 plus traffic from Waco headed south. Momentarily the southbound will be underway from Hill Yard.

Lower left and below . . . While GE 5111 sorts cars in Hill Yard, through trains roll by on the main. The 9404, built in 1975, is the highest-numbered of the hundreds of SP/SSW SD45T-2's. On a March 1982 morning, the last unit of twenty-five in class EF636C-9, locomotives 9380-9404, eases by the 5111. The primary function of the small yard is to hold cars to and from Waco.

Above . . . The Cotton Belt's influence still persists on the east side of Corsicana. Strangely enough, Cotton Belt Drive is not the road which parallels Hill Yard, but it does run perpendicularly north from the yard.

Above and right . . . Fuel-efficient, twelve cylinder prime movers power the fifteen B23-7's (5100-5114) which came from GE in 1980. The roof-top cab air-conditioner is not needed on this mild December afternoon in 1981 as 5106 switches the east end of Hill Yard. Although assigned to the SP proper, the 5100's work on the Cotton Belt with a maintenance base of Pine Bluff, Arkansas.

SP
Southern Pacific
Golden Pig Service

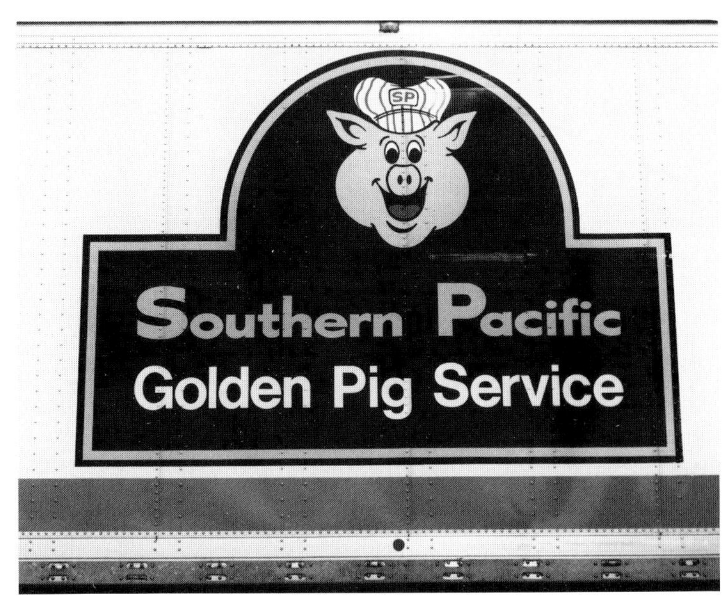

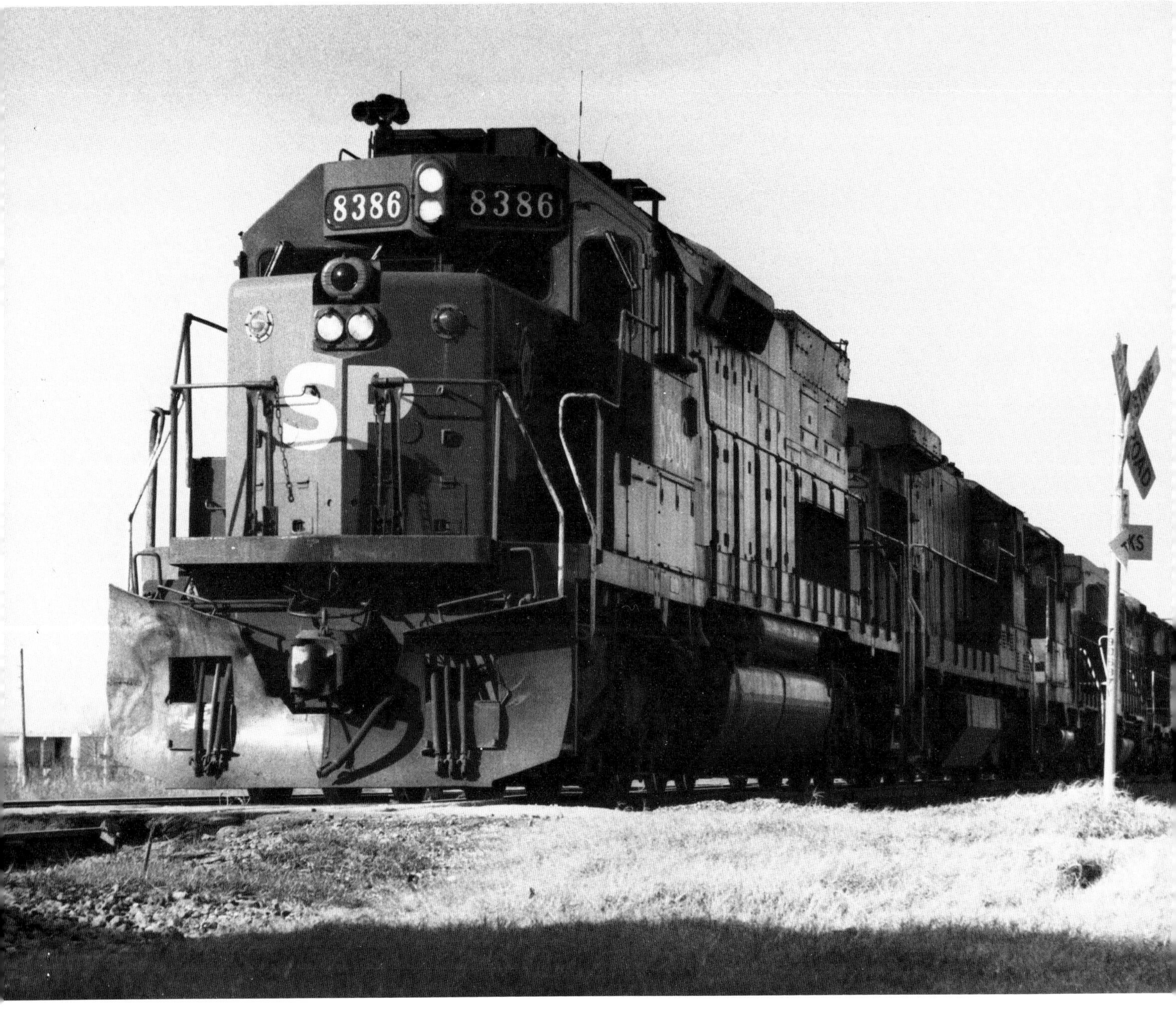
8386
8386
SP

Left . . . On the day after Christmas 1981, SP Locotrol remote unit 8386 finds itself on the point of a northbound piggybacker leaving Corsicana. Low-numbered 8300's are masters and units numbered upward from 8350 are remotes. Locotrol-equipped 8300-series SD40T-2's came to the SP in 1974, 1978 and 1979, complete with the nine-foot-long extended nose.

Right . . . The cowboy-hatted engineer in charge of some Hill Yard switching surveys the scene from his perch in GE 5103. Eventually the B23-7 will be the trailing unit on the Waco train, poised on the adjacent track in the distance. In Texas, men still wear their Stetsons in restaurants, in locomotive cabs, and while driving their pickups (or Cadillacs). Actually the pickup is the accepted vehicle, even if you don't plan on carrying anything larger than a six-pack.

9328
9328
SOUTHERN PACIFIC
9328
SP

Four of SP's stable of six-axle EMD's, 9328-9185-8323-9019, lead an outbound Cotton Belt freight past Tower 72 and across the Fort Worth & Denver (BN) crossing in Corsicana, *above*. A mile or so further east, the 9328 is passing Hill Yard, *left*. Beginning with locomotives 8800-8844 in 1966, SD45 purchases ran upward through the 8900's and 9000's to 9156 by 1971. Then, from 1972 through 1975, SD45T-2's followed, numbered 9157 through 9404. The "T" stands for "Tunnel" motor, an SP innovation to reduce the risk of overheating in tunnels by lowering the air intakes (at the rear of the unit) to draw cooler air from nearer the ground.

Above . . . An August 1980 graduate of the Sacramento-upgrade program, SD40E 7314, switches some cars in front of the Corsicana depot in December 1981. Trackage north to Ennis and Dallas is to the left of the ex-8416.

Below . . . SD40T-2 8539 rounds the curve off Southern Pacific trackage and onto SSW rails toward Arkansas in October 1982. The present Corsicana depot is visible between GP's 3368-3356, which will be working the Waco Branch train, and Cotton Belt SW1500 2580.

A quartet of GP35's and older Geeps bangs across the Waco Branch diamond with tonnage bound initially for the yard in Ennis, twenty-two miles north. Leading the parade is the 6623, a 1965-built '35, delivered as SP 7726.

A six-unit mix of GE's and EMD's from the SP and the Burlington Northern drag the empty hoppers of a SATX unit coal train north through Corsicana. Observing U33C 8719 (note the non-block SP letters on the short hood) are BN SD40-2 7265, *above*, and block-lettered SP GP9E 3732, *below*.

Above . . . Spending a mid-December 1982 week in Corsicana as the local switch engine, is MP15AC 2753, EMD class of 1975. The caboose belongs to a Cotton Belt-bound freight up from Hearne on the SP.

Left . . . Multi-headlighted diesel power has been an intriguing part of SP/SSW history. EMD 9309 has the standard share for SD45T-2's as it waits to depart Corsicana in February 1982.

GP35 6617 stands aside as a quintet of six-axle tunnel motors hauls the *BSM* off the Cotton Belt and south on the SP in December 1982.

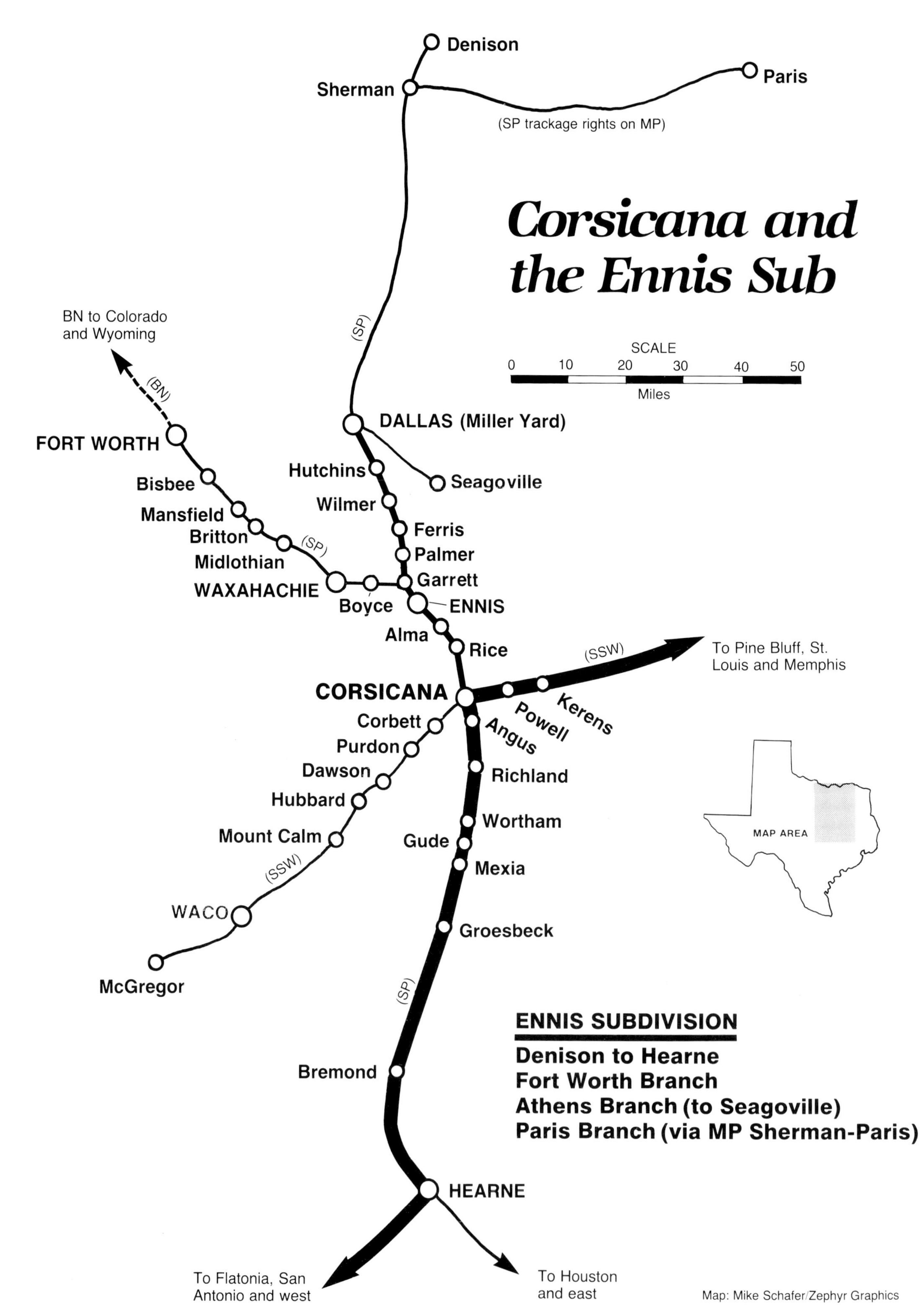

Corsicana and the Ennis Sub
Denison
Sherman
Paris
(SP trackage rights on MP)
(SP)
BN to Colorado and Wyoming
(BN)
SCALE
0
10
20
30
40
50
Miles
DALLAS (Miller Yard)
FORT WORTH
Hutchins
Seagoville
Bisbee
Wilmer
Mansfield
Ferris
Britton
Palmer
Midlothian
(SP)
Garrett
WAXAHACHIE
Boyce
ENNIS
Alma
Rice
(SSW)
To Pine Bluff, St. Louis and Memphis
CORSICANA
Kerens
Powell
Corbett
Angus
Purdon
Dawson
Richland
Hubbard
Wortham
Mount Calm
Gude
Mexia
MAP AREA
(SSW)
WACO
Groesbeck
McGregor
(SP)
ENNIS SUBDIVISION
Denison to Hearne
Fort Worth Branch
Athens Branch (to Seagoville)
Paris Branch (via MP Sherman-Paris)
Bremond
HEARNE
To Flatonia, San Antonio and west
To Houston and east
Map: Mike Schafer/Zephyr Graphics

Burlington Northern power, including the 5009, a September 1979 C30-7 from General Electric, waits for a call north to Ennis, Fort Worth and home rails. Cruising by on the Cotton Belt connection is SD40T-2 8538, leading the *BSM* toward Hearne. With above-freezing weather and fewer than 120 cars, *BSM* is authorized 70 miles per hour between Corsicana and Hearne on the former H&TC.

ROAD
8518
SP
RAIL ROAD
CROSSING

Left . . . Four EMD's, and a lone GE, urge a southbound freight (but timetable eastward) away from a Corsicana crew change. Following the 1966-68 SD40's (8400-8488), SD40T-2's 8489-8533 were built for the SP in 1978 and 8534-8573 in 1979. EMD 8518 is three years old on this December 1981 noon. Since 8585 and upward were occupied by U33C's, other SD40T-2's were numbered 8300's (the Locotrol units) and then in the 8200's.

Below . . . SSW 7657-7656 are grinding out of Corsicana on sanded rail with a piggybacker in December 1981. By the time the caboose passes, speed will be up to 35 and increasing.

CORSICANA-GROESBECK

Left . . . SD45 9074 and mates are emerging from the north end of the siding at Angus, Texas in the wake of a meet with a pig train rapidly disappearing in the distance. October 1982.

Below . . . Six weeks later, at the south end of Angus siding, 8526 roars by on welded-rail at 60 or more. When the dust settles, the westward freight, in the hole, will continue to Corsicana.

Right . . . The engineers in EMD 8518 and EMD 7657 (pages 30 and 31) are looking directly at this freight speed limit sign as they depart Corsicana headed south on the former Houston & Texas Central.

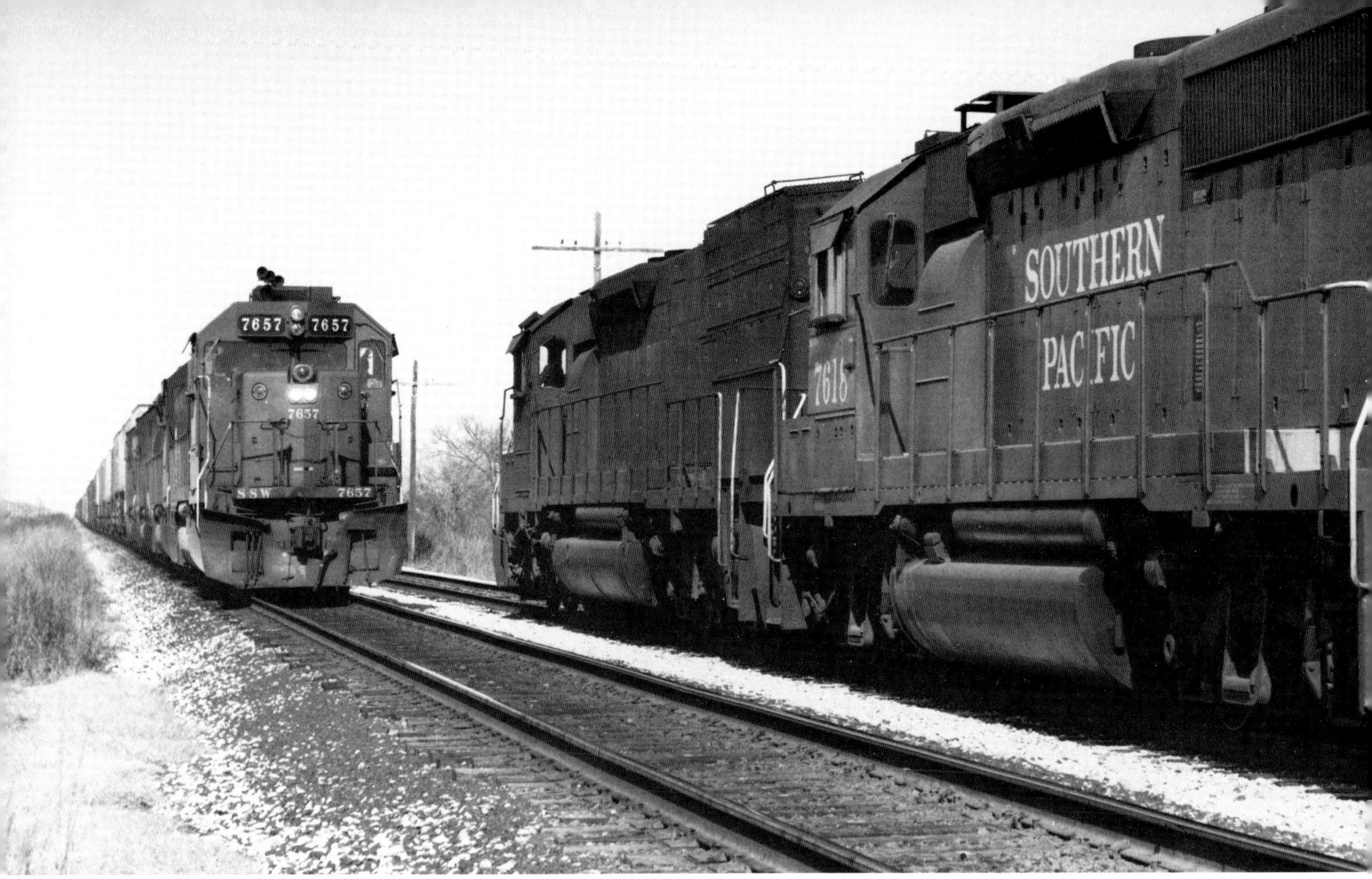
7657
7657
7637
SSW
7637
SOUTHERN
PACIFIC
7618

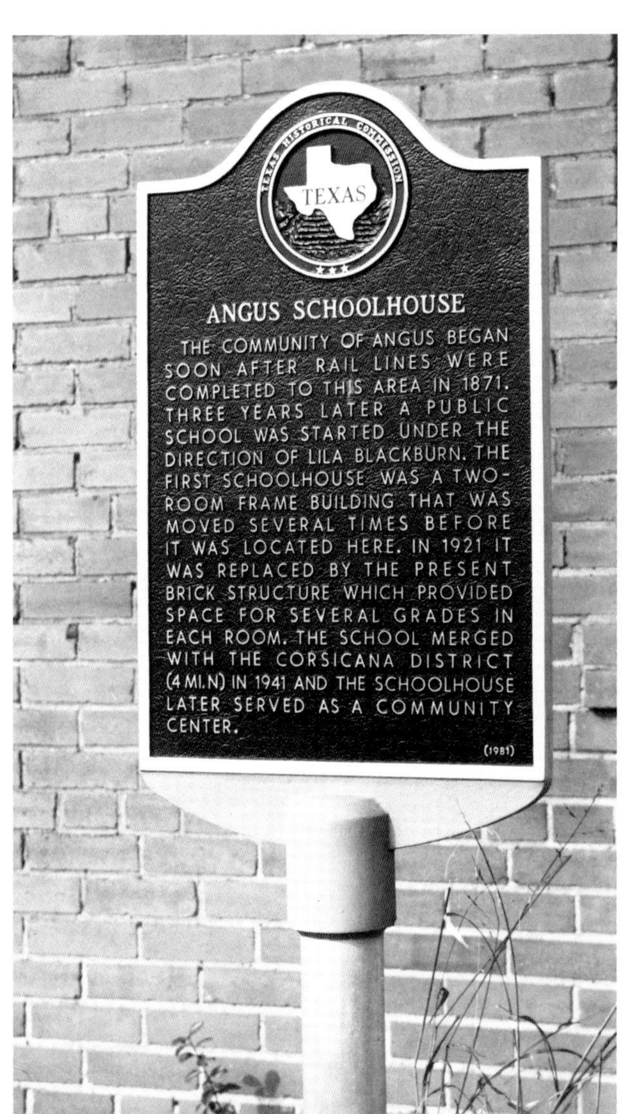

TEXAS HISTORICAL COMMISSION
TEXAS
ANGUS SCHOOLHOUSE
THE COMMUNITY OF ANGUS BEGAN SOON AFTER RAIL LINES WERE COMPLETED TO THIS AREA IN 1871. THREE YEARS LATER A PUBLIC SCHOOL WAS STARTED UNDER THE DIRECTION OF LILA BLACKBURN. THE FIRST SCHOOLHOUSE WAS A TWO-ROOM FRAME BUILDING THAT WAS MOVED SEVERAL TIMES BEFORE IT WAS LOCATED HERE. IN 1921 IT WAS REPLACED BY THE PRESENT BRICK STRUCTURE WHICH PROVIDED SPACE FOR SEVERAL GRADES IN EACH ROOM. THE SCHOOL MERGED WITH THE CORSICANA DISTRICT (4 MI. N) IN 1941 AND THE SCHOOLHOUSE LATER SERVED AS A COMMUNITY CENTER.
(1981)

Above . . . Tunnel motor 9384, from the last order of '45T-2's in 1975, races south across the Texas countryside approaching Angus.

Upper left . . . A running meet is underway at Angus in December 1981. To the left, SSW 7657 is approaching at better than 50 miles per hour. On the right, already safely tucked in the passing track, SD45T-2 9291 and GP40-2 7618 lead the power mix rolling north toward Corsicana.

Lower left . . . As noted on the historical marker, Angus dates to the coming of the railroad in 1871.

Above . . . GP40-2's 7629 and 7628 accelerate past Angus on the rolling speedway between Corsicana and Hearne in December 1982. In the 1950's, Numbers 13 and 14, the *Sunbeams*, were alloted 265 minutes for their 264 mile journey between Dallas and Houston on this route.

Above . . . Water from December rains (and a little snow too) provides a wavy reflection of GP20E 4139 as the Sacramento upgrade eases up the Angus siding in 1982. The mixed bag of power also includes MP15AC 2755, SD45T-2 9255 and SD40E 7351. EMD 4139 entered the world as Cotton Belt 809 in January 1961 and was reworked in 1975.

Lower left . . . Its opponent has not yet arrived, so the quartet of GP40-2's led by 7625 is only creeping toward the Angus, Texas siding on a February afternoon in 1982. The 7625 is geographically headed south, leads a timetable eastward train, but is enroute west across Texas. (Dallas to Hearne and Houston on the old H&TC is eastward; beyond Hearne toward San Antonio this train will again be westward.)

Southern Pacific
Transportation Company

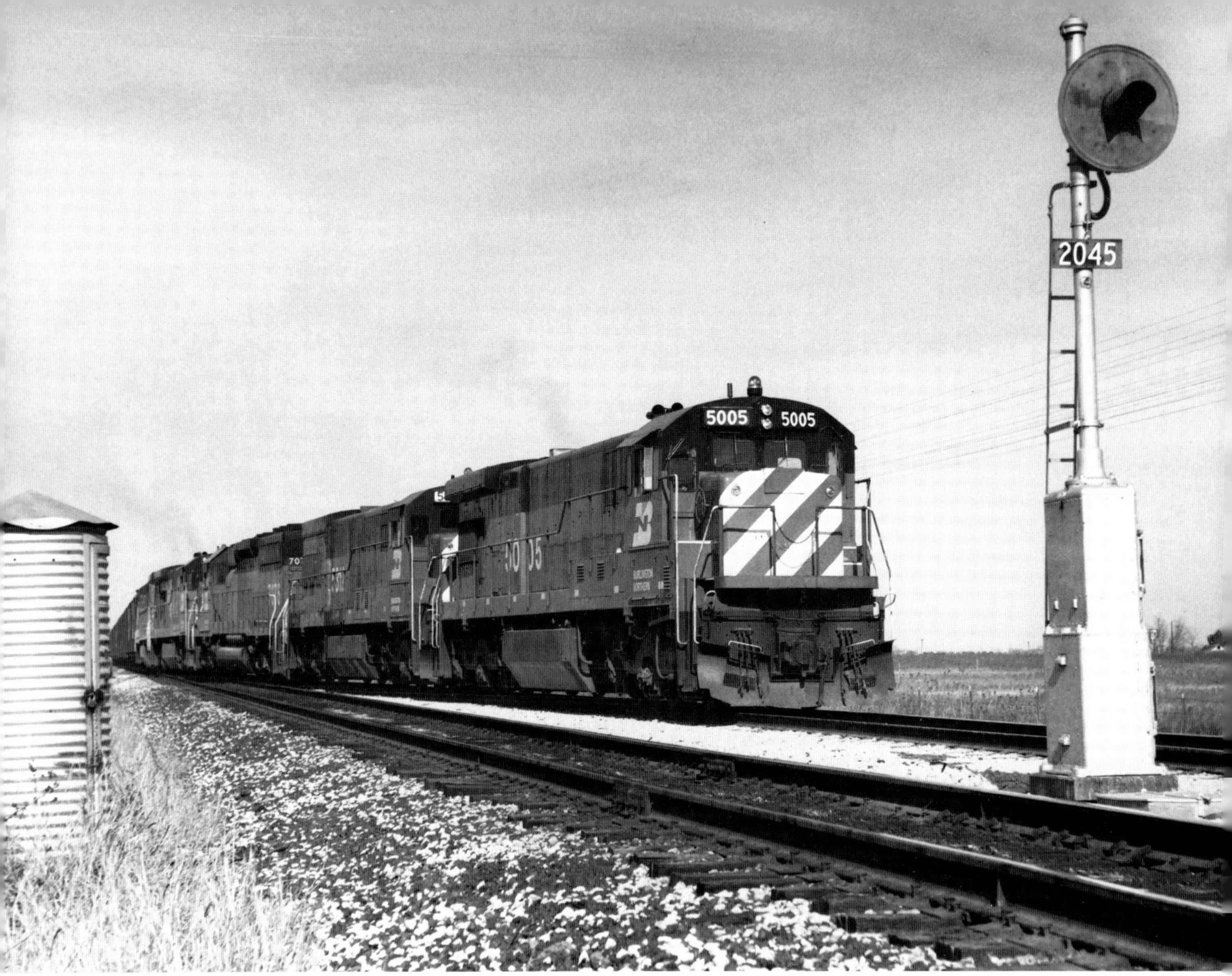

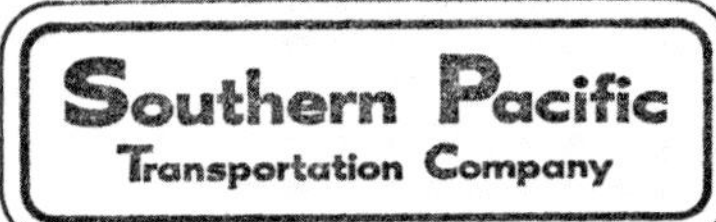

TRAIN ORDER No. 966 DATE DEC 29 1981

To C&E

EXTRA 8975 WEST

Station GROESBECK

EXTRA 8975 WEST MEET

EXTRA BN 5005 EAST

AT ANGUS

An order consummated! Burlington Northern C30-7 5005 heads into Angus siding, trailed by mates 5312-7076-5014-5084 and a SATX coal train, *above*. A short time later SSW 8975 drifts by on the main with a Corsicana-bound freight, *upper right*. After the path is clear, BN 5005 East is again underway with its coal loads, bound for a San Antonio power plant, *lower right photograph*.

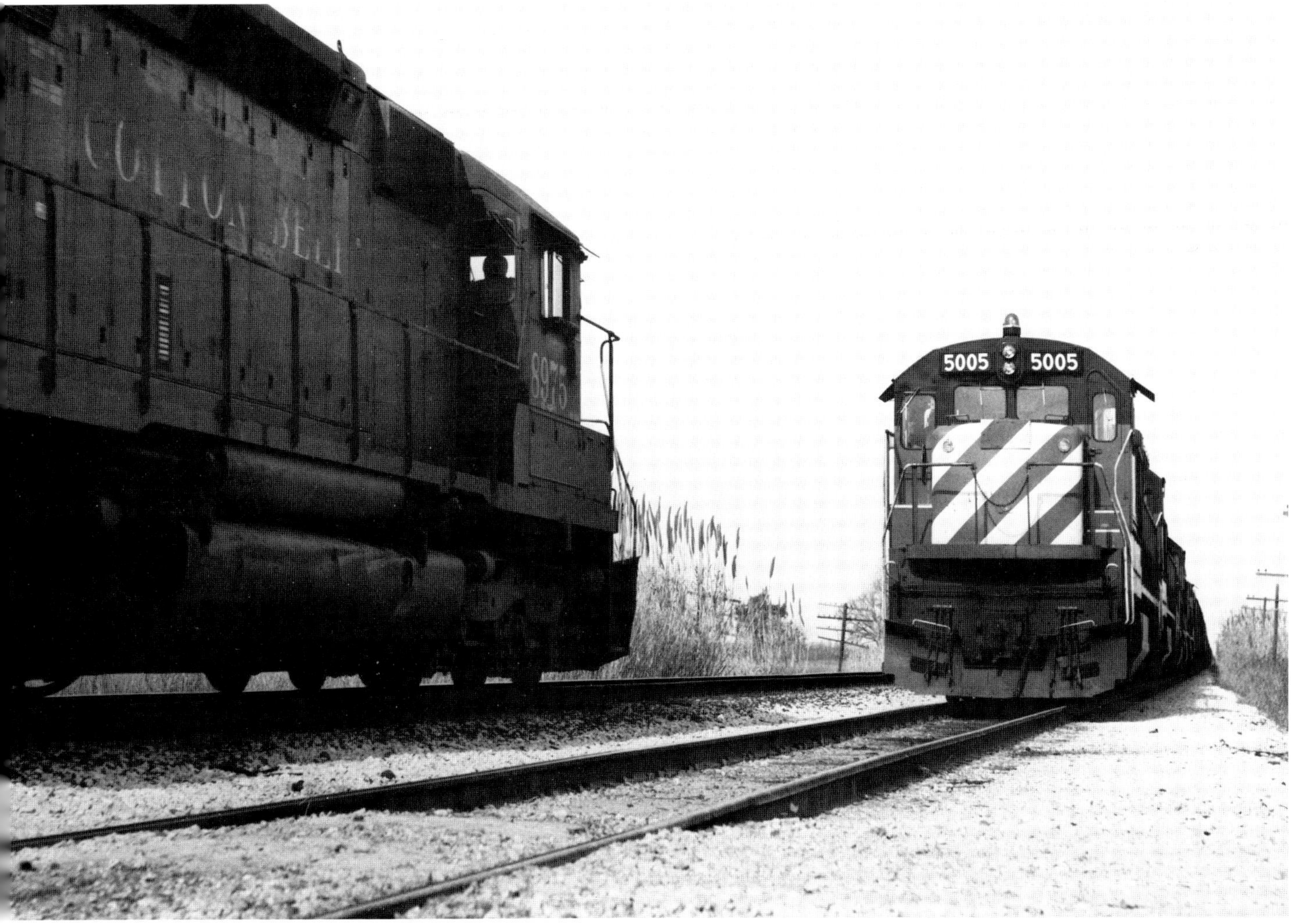
5005 5005
8975

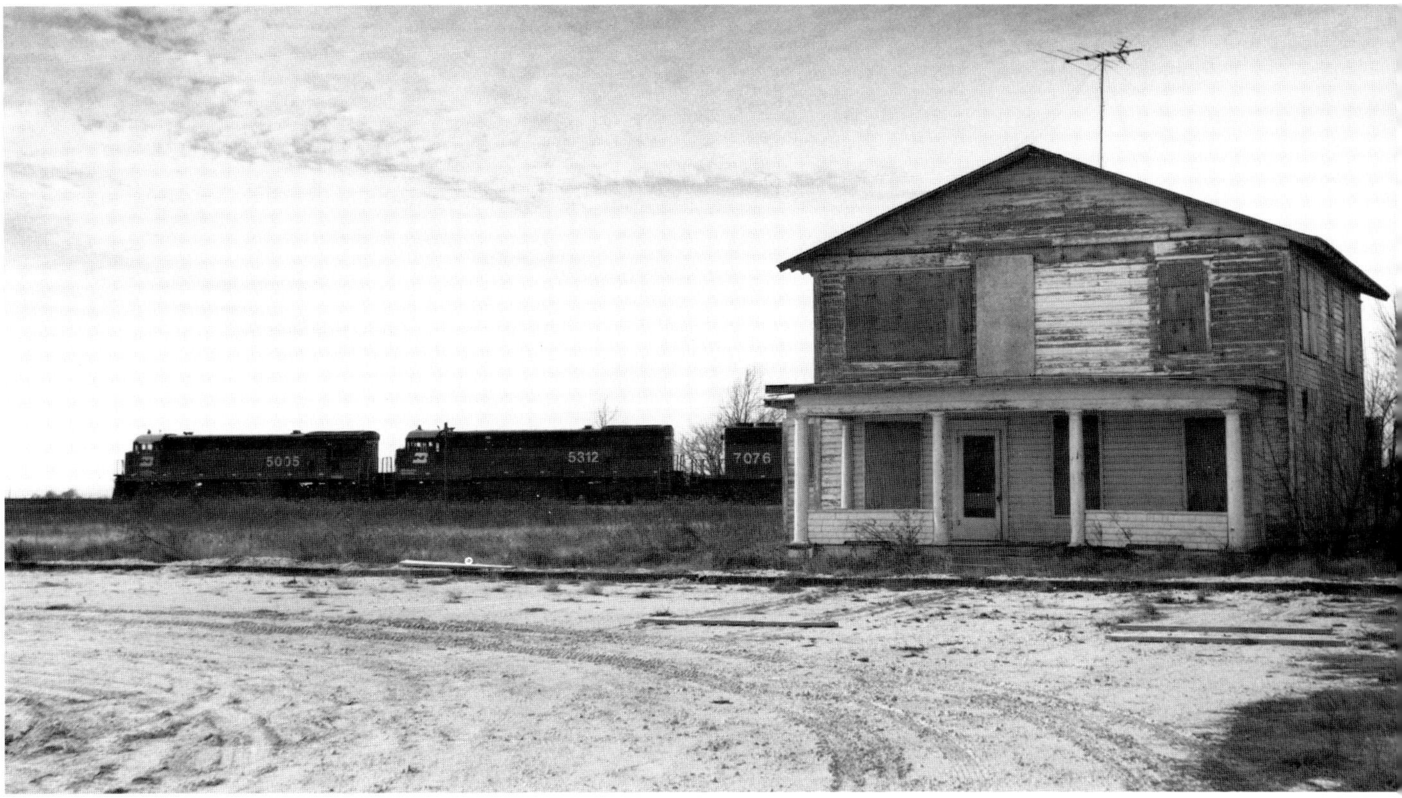
5005
5312
7076

Left . . . Signal 2045 guards the main at the north end of Angus siding.

Above . . . On a magnificent November afternoon, red and gray 8546 hustles south between Angus and Richland. Five miles of steady downhill running starts at Angus, so speeds reach the posted limits here.

Above and below . . . Track crews have cleared away the brush which normally lines Texas railroads as Cotton Belt 7638 and SP 7625 command a pair of south-facing freights approaching the track gangs in February 1982.

One of Espee's hundreds of bay-windowed cabooses trails an auto-rack train toward Richland and Hearne in 1981.

Another SATX coal train pounds south beyond Angus behind the efforts of SD40-2 7083 and a quartet of BN GE's. The nearly 14,000-ton trains originate at a mine near Gillette, Wyoming on the Burlington Northern and are turned over to the SP in Fort Worth. Mixed SP and BN locomotive consists have given way to solid BN power on these SATX trains since about mid-1982. SP U33C's used in coal service were sidetracked at that time, apparently due to high fuel/maintenance costs.

7629
7629
7629
COTTON BELT
COTTON BELT
7627

SOUTHERN
PACIFIC
7624

Above . . . Near Wortham, Texas, the 7638 fronts a fast-running *BSM* in December 1982.

Upper left . . . Power for many of the freights working on and off the SSW at Corsicana in 1982 was a GP40-2 quartet, often chained together in consecutively numbered sets. Such is the case here as Cotton Belt 7629-28-27-26, with help from SD 9374, pull south near Richland.

Lower left . . . Another four-unit lashup of GP40-2's leans into a super-elevated curve north of Richland.

In January 1982, a pair of EMD's, 7306 and 9324, lead the four-unit power team negotiating one of several wood trestles located about three miles north of Wortham. SD40E 7306 was upgraded at Sacramento in 1980 from SD40 8434, built in 1966.

Above . . . The low level air intake at the rear of SD45T-2's is readily visible as tunnel motor 9288 works south at Angus siding. Forty-one of the EF636's built at La Grange in 1973 were assigned to the Cotton Belt, road numbers 9261-9301.

Below . . . SD40E 7306 charges through Richland, Texas on the SP's high-speed, single-track main between Corsicana and Hearne.

Above . . . SATX coal loads roll toward San Antonio under darkening skies heralding the approach of a winter storm.

Left . . . Interestingly numbered GP35 6666 spends a quiet Sunday morning in Mexia (pronounced meh-HAY'-ah), Texas. Sister 6623 is in command of a Hearne-Ennis freight. Both engines are among the dozens of GP35's which came from EMD during 1965.

Above right . . . This unusual approach sign guards the SP main in Groesbeck, Texas.

Above left . . . Although the great herds no longer roam this part of Texas, there is still a fair amount of "posted" land devoted to cattle grazing.

Right . . . SD45 8954 leads the curving charge up the short, but steep, 1.5 percent eastward grade which crests right in downtown Mexia. A heavy or underpowered train could be slowed down pretty well by the climb, but on this November 1981 afternoon the EMD/GE mix is moving right along.

8595
8954
8954
SP

WACO BRANCH

Above . . . In December 1981, one hundred years after the Texas & St. Louis of Texas laid the rails to Waco, Southern Pacific's 2883 curves through Corsicana with a Waco-bound freight. Trailing the Geep are GP35 6615, GP9E 3389 and GP35 6514. The 2883 is one of the twenty GP9's which were built at EMD with a low-nose, and was originally the 5883.

Upper right . . . GP9E's 3368 and 3356 are working from the east end of Hill Yard as they make up the Waco train. Shortly after 1:00 P.M., the same pair will depart for Waco from the west end.

Lower right . . . Heading #155 through Corsicana toward Waco is GP35 6680 on a mostly cloudy April 1982 afternoon. The 6680 is the next-to-last GP35 built for the SP. Along with 6681, these two 1965-built GP35's were diverted to the Cotton Belt where they first carried numbers 780 and 781.

SOUTHERN PACIFIC
3368
3368
3368
SP

Above . . . On a super January 1983 day, the assigned units for the Waco train were facing the wrong way, so GP30 5006 and GP9E 3351 are being turned on the wye at the west edge of Corsicana.

Below . . . With their act back together, the duo sets forth on the jaunt to Waco.

Above . . . Upgraded GP9E 3810, assigned to the Cotton Belt, and B23-7 5103 roll toward Waco in December 1982. The GP9 came from La Grange in 1957 as SSW 824 and was released from upgrading in June 1975.

Below . . . Two more upgrades, GP9E's 3356 and 3368, are just beginning their Waco excursion as they pull through Corsicana.

Below . . . Only eighteen GP30's were ordered by the Southern Pacific and ten of these were assigned to the SSW. Originally 750-759, the Cotton Belt units were renumbered 5000-5009. In January 1983, it is hard to believe these early second-generation locomotives already are twenty years old. Although the SP has been rebuilding or upgrading everything from slugs and switchers to SD45's, there apparently are no plans to rework the 1963-built GP30's. So major mechanical problems may be the downfall of these units. Here, the 5006 leads the way toward Waco, nearing Corbet.

Right . . . Cotton Belt 6680 drifts through steamy, backwoods country only a few miles out of Corsicana.

COTTON
6680

RAIL
CROSSING

Above . . . Very few cabooses on the entire SP/SSW roster are not of the bay window variety. Among them are SSW 1-25, class C-40-8, built by International Car in 1959. Here, wide-vision cupola SSW 4 trails a Waco-bound freight through Corbet, Texas, in 1983.

Below and left . . . A quartet of older units, GP35 6514, GP9E 3389, GP35 6615 and GP9E 2883, cut through the brushy, sparsely-populated countryside about halfway between Corsicana and Corbet in January 1982.

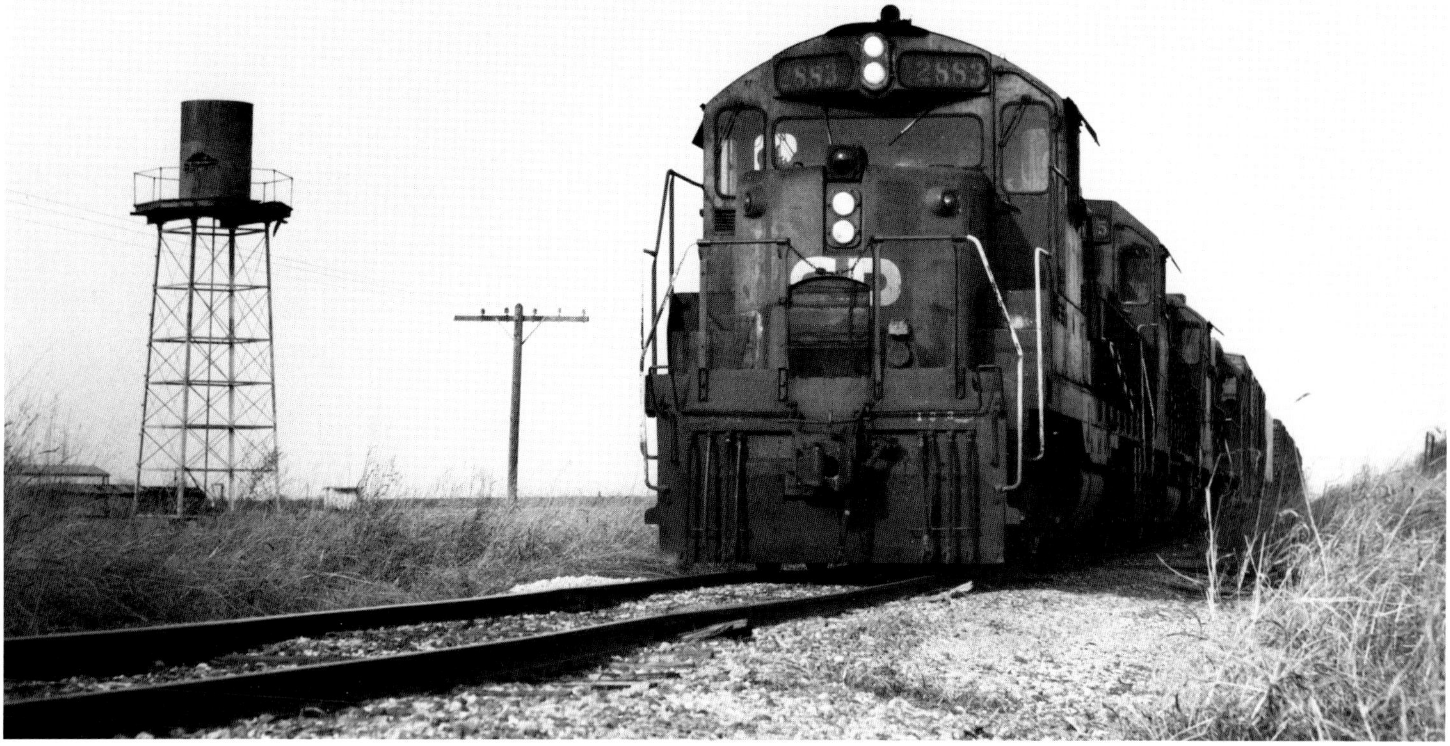

2883
COTTON BELT 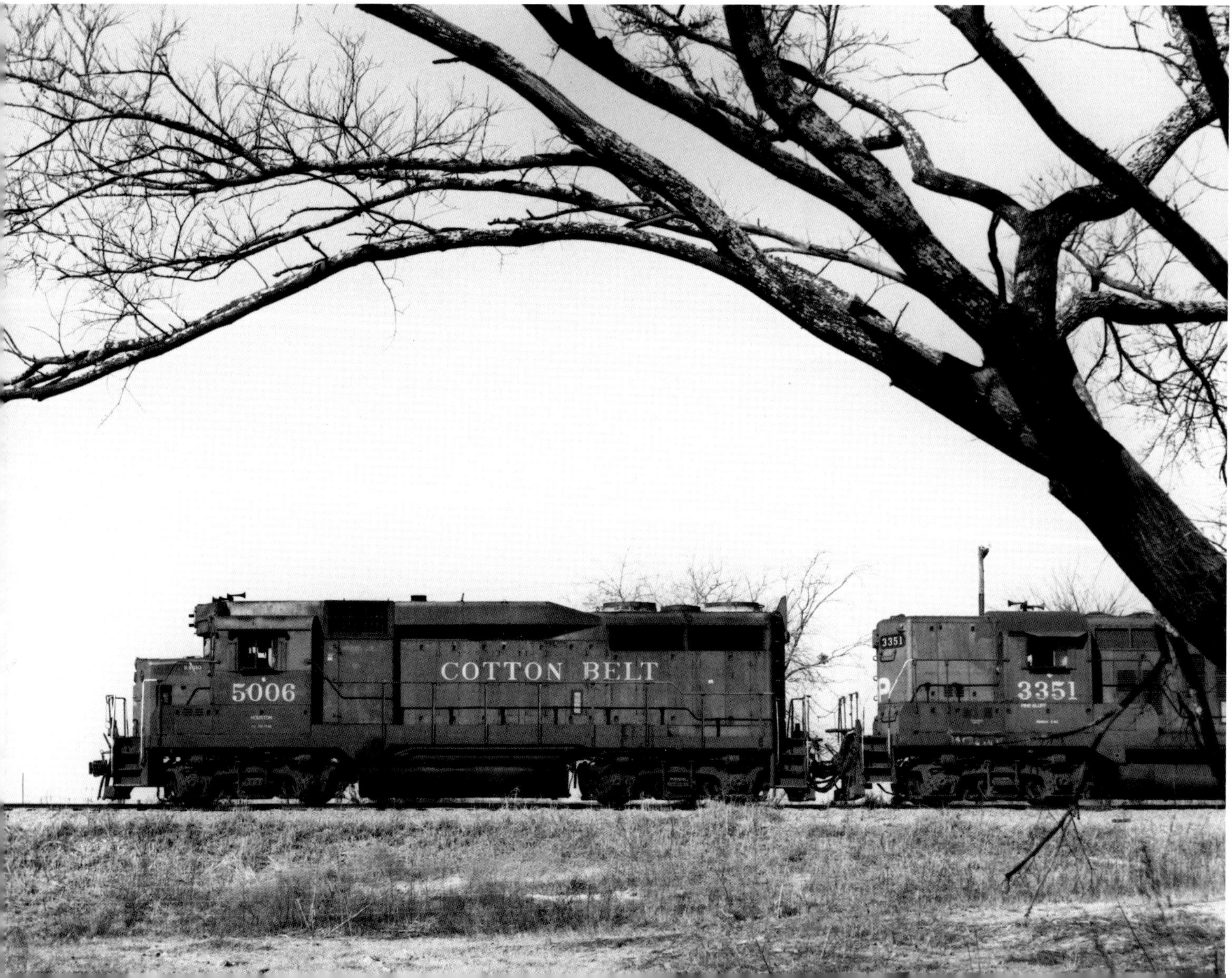
5006
3351

Upper left . . . Late on a December afternoon, a weakening sun casts its nearly horizontal light on Waco-bound GP9E 2883 in Corbet.

Lower left . . . GP30 5006 is framed by the bare branches of a large tree, just west of Corsicana.

Below . . . About a mile west of Corbet (down a one-lane paved road), the Waco Branch crosses a long, wood trestle. GP35 6514 drifts across this trestle over a normally dry flood plain (except after a Texas-sized thunderstorm!).

Above . . . The Waco train curves along west of Dawson, Texas.

Below . . . Dawson's Christmas Tree is still decorated on this January 1983 afternoon as the 5006 chants toward Waco.

Lower right . . . 5006 and 3351 labor up a slight grade into Hubbard, 26.9 miles out of Corsicana. At the time of this writing, track conditions required speeds of less than 20 miles per hour on the Waco Branch, quite a contrast to the 60-70 mph mainline limits.

Upper right . . . Geeps 3356 and 3368 ease their short train across the wood trestle west of Corbet.

3356
3356
SP

Above . . . It's about 2:30 in the afternoon as the 5006 and 3351 clatter along about two miles northeast of Hubbard, Texas, on their journey to Waco.

Right . . . A light sprinkle occasionally falls from the sun/cloud mix as the 3356 grinds into Hubbard in October 1982. Depending on business, the run to Waco may occur daily (except Sunday), or may be Tuesday, Thursday and Saturday only. Normal departure time is the early afternoon. The return trip usually reaches Corsicana before 9:00 A.M.

3356
PINE BLUFF

CORSICANA-ENNIS

Above . . . At the south end of its twenty-two mile sprint from Ennis, upgraded GP9E 3777 hustles into Corsicana with a short local in April 1982. It already has been well over six years since the 3777 went through the Sacramento program.

Above . . . A steady rain is falling as a lashup of six Burlington Northern and SP diesels churns south through Alma, Texas, with the SATX coal train. BN U30C 5352 was constructed by General Electric in late 1972 so is approaching a decade of duty on this rainy April 1982 day.

South of Rice, BN 5352 and the SATX loads continue on their wet journey toward Corsicana, Hearne and San Antonio. From late February or early March through April into May is the most cloudy, showery time of the year in this area of Texas. Late May until mid-September brings heat, lots of it, with afternoon temperatures almost always between 93 and 103 degrees, but plenty of sunshine. The best months for a combination of warm weather and sun is from late September until around Christmas. January and February have a number of mild, sunny days, but also rainy periods and usually a couple of ice storms per winter. Snow is rare south from Dallas and Fort Worth, but much more likely west and north.

SATX
3030
RAILROAD
CROSSING

SOUTHERN
PACIFIC
1960
RAILROAD
CROSSING

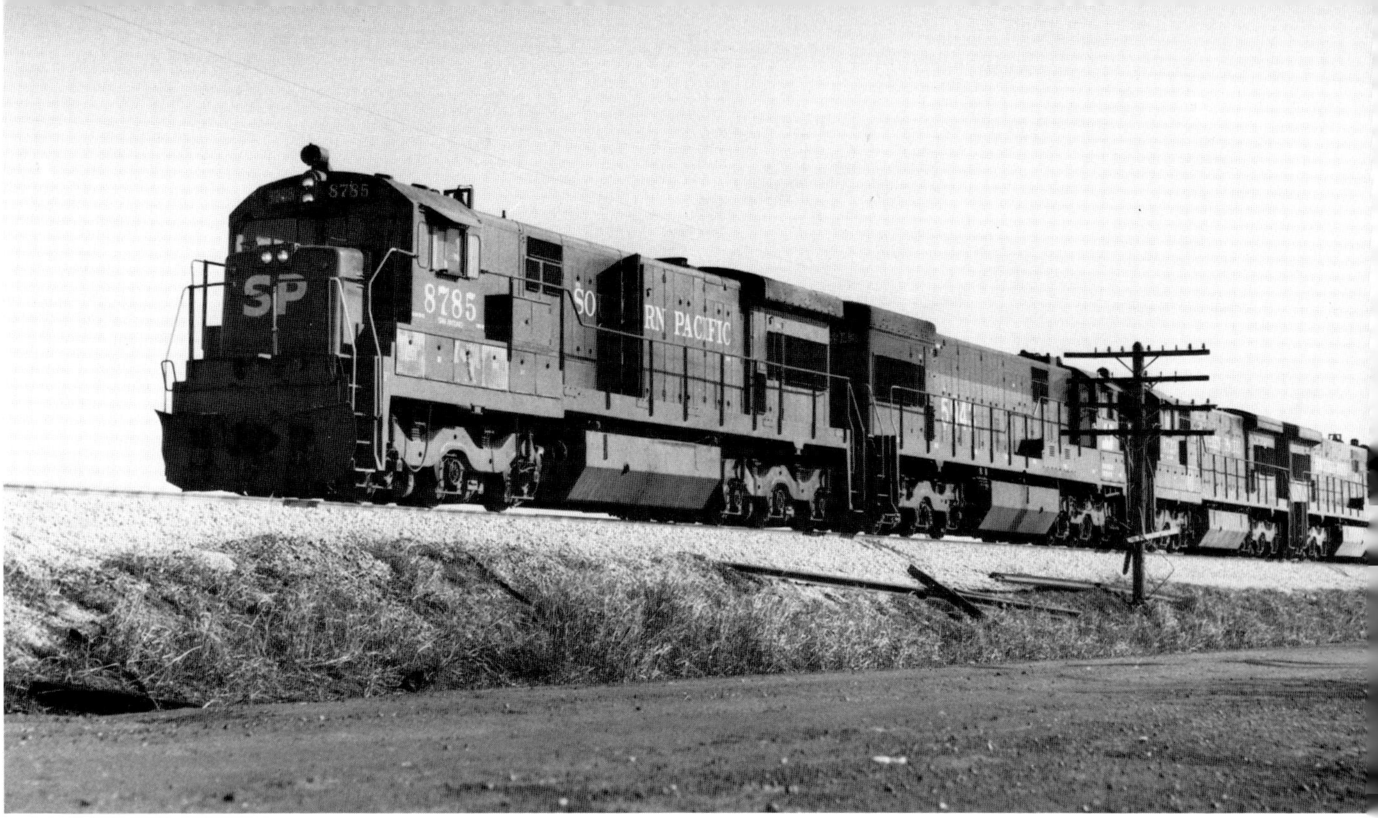
8785
SP
8785

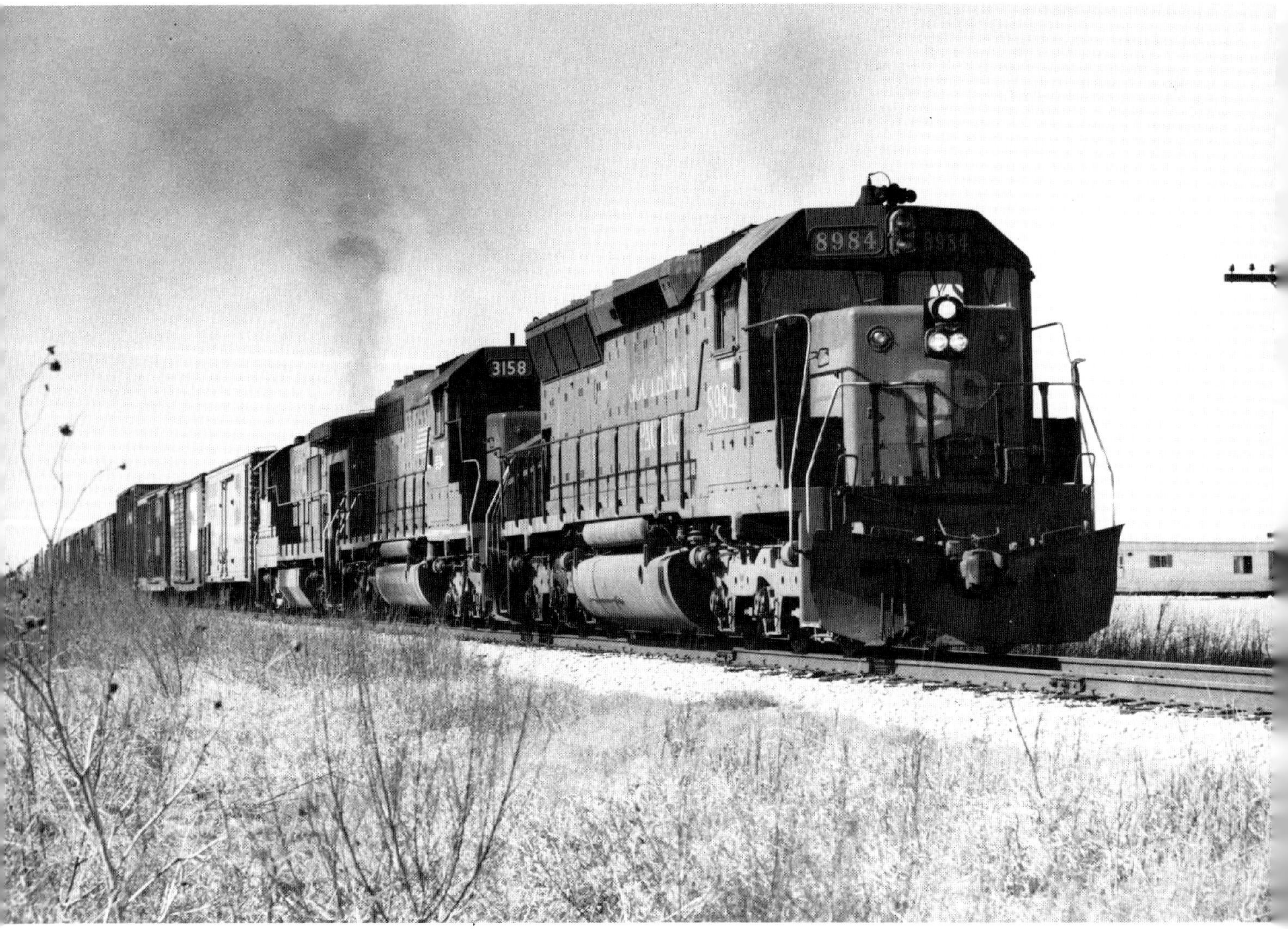
8984
8984
3158

Left . . . An all-GE team of six SP and BN units pulls the empty hoppers of a SATX train north through Rice in December 1981. SP 8785 was the last of the eighteen U33C's built for the railroad in 1973. Locomotives 8786-8796 came in 1974, followed by 8585-8599 in 1975.

Lower left . . . Missouri Pacific SD40-2 3158 is sandwiched between SP 8984 and SSW 7789 on an eastward freight between Ennis and Rice in February 1982.

Below . . . On December 22nd, the sun rises directly above the SP main extending southeast out of Ennis. After the early morning low clouds clear, this will be a fine day.

Above . . . On this cold January morning in Rice, Texas, we're in the cab of 8719 as the Ennis-Corsicana local approaches behind Cotton Belt 7786 and 8324. Stretched out behind SP 8719 are the SATX coal empties. GE 7786 is one of the twenty-six B30-7's delivered in 1980.

Upper right . . . A short westward freight creeps up the Rice passing siding as BN 5086 thunders by with a January 1983 SATX coal train. Cascade green 5086 is a 1980 version of GE's C30-7.

Lower right . . . This time it's EMD's on the point of a SATX holding the main at Rice. SD40-2 7254 was built at approximately the same time as GE 5086, early 1980. Three cabooses, SP 4062, SSW 73 and SSW 78, trail the local up the siding. The temperature is already nudging 90 degrees on a July 1982 morning.

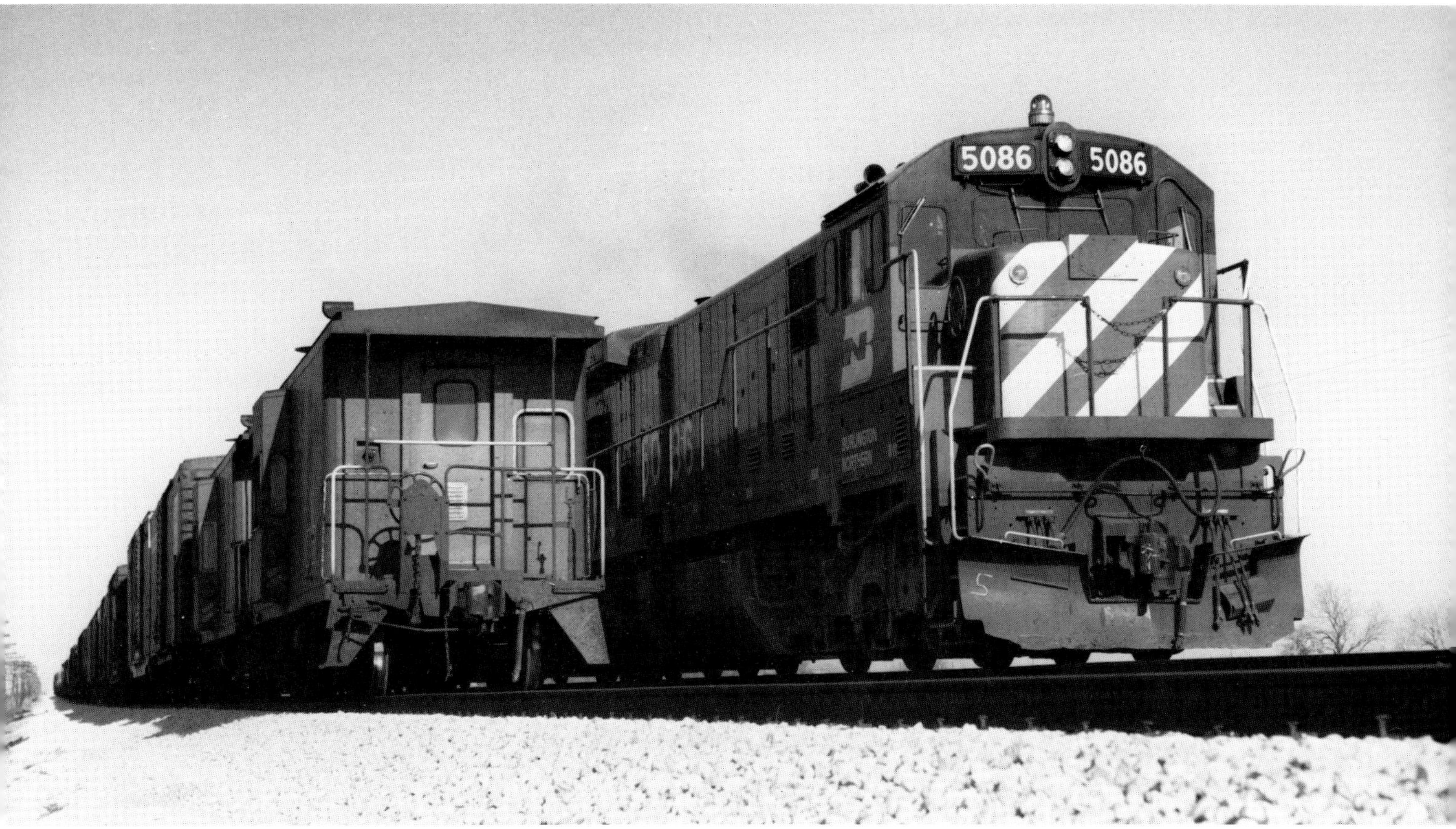
5086
5086

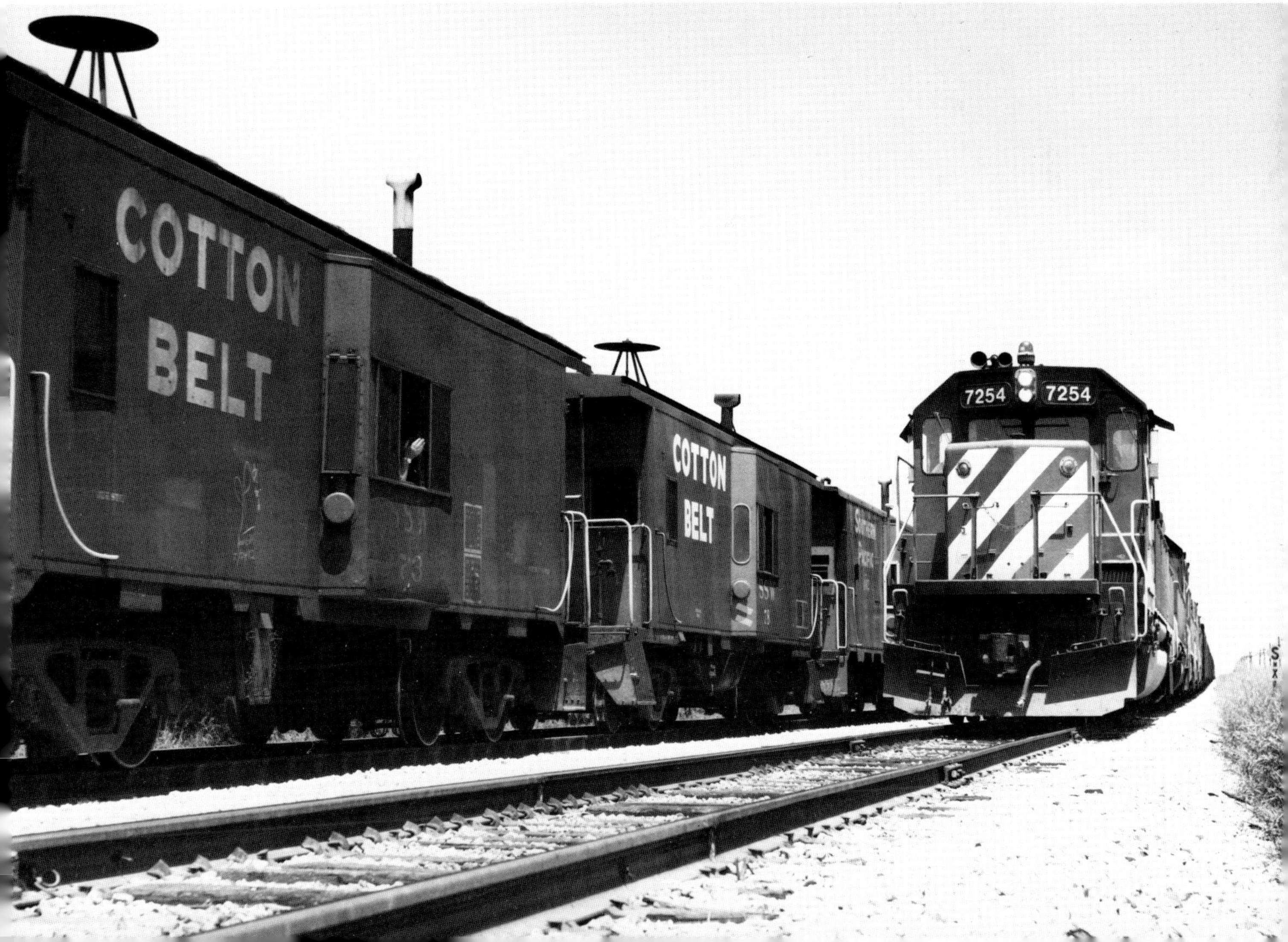
COTTON
BELT
COTTON
BELT
7254
7254

Nearer the camera, GE 8699 stands at the north end of the Ennis yard, ready to roll with the SATX empties. Entering the yard is sister U33C 8669 with a piggyback train. December 1981.

Above . . . An ex-Frisco caboose (BN 11517) and a Rock Island commuter coach are temporarily parked on an industrial siding at Ennis, Texas, before delivery to a private owner. Accelerating south toward Corsicana on a cloudy April 1982 day are BN 5819 and 7803, point units for a SATX coal train.

Below . . . Earlier the same gray morning, before the coal train left Ennis, SD45T-2 9271 led the arriving power cutting off an inbound.

FORT WORTH BRANCH

Above . . . Shortly after sunrise on a crisp December 1981 morning, U33C 8751 leads the five GE's battling out of Ennis for Forth Worth with SATX coal empties.

Lower left . . . A few minutes earlier, the SATX loads pulled to a stop in the Ennis yard. In the early 1980's, the loaded trains normally reached Ennis from Fort Worth very early and were down to Corsicana between 10:30 and noon, to follow *BSM* (and any other hotshot) south. (With *BSM* shifting to a Kansas City routing, this has changed to later in the day.) The empties were not so predictable, but morning out of Ennis was more likely than afternoon.

Upper left . . . The Fort Worth Branch actually begins here at Garrett, 1.9 miles north of Ennis. BN 7179, aided by SP U33C 8725 and B30-7 7849, is approaching the junction with the Ennis-Dallas main (in the foreground) with a train from Fort Worth. SD40-2 7179 was built in Canada in 1979 when La Grange was overbooked with orders. Not much over 18 months later, locomotive construction was down to a trickle because of the business downturn.

SP 8699, and a mixed EMD/GE lashup of seven BN units, swings away from the Dallas line at Garrett with SATX empties, *below*. Five miles west, the power is working up out of a sag approaching Boyce, *above*.

Above . . . Eastward-moving loaded SATX coal trains receive a helper between Forth Worth and Ennis. There are several fairly stiff grades on this route, in particular the climb into Midlothian which crests right at the AT&SF crossing. The shoving is over for BN 7266 now, and the SD40-2 is just along for the ride as SATX drifts into Ennis Yard.

Below . . . In 1982 an auto-unloading area was opened in Midlothian. The Ennis to Fort Worth local turn now regularly has some head-end auto racks to set out. On January 24, 1983, three older SP units (6582, 3384 and 6620) are aided by BN 5070 as the turn rolls west out of Garrett.

Six GE's are on the point, and one SD40-2 is pushing (BN 7266 on the previous page), as a SATX coal train negotiates the roller-coaster profile of the Fort Worth Branch between Waxahachie and Garrett. From front to back on this not-yet-hot May 1982 morning the GE's are BN 5120 and 5383, SP 8711-8744-8717, and BN 5318. Southern Pacific's decade-old 8700-series U33C's were removed from service shortly afterward (along with all other six-axle GE's).

5120
5120
BURLINGTON
NORTHERN

5108
SOUTHERN
PACIFIC

Left . . . Two miles west of Boyce, Texas, GP35 6547 and B23-7 5108 head a Fort Worth-bound turn from Ennis on a darkening December morning.

Above and below . . . A rare combination of power, SD45T-2's 9255 and 9253 plus BN 5009, is in charge of the empty hoppers of a SATX train near Boyce in early December 1982. Solid BN engine consists were standard on the SATX trains at this time.

Between the curves at Garrett and Waxahachie, on the old Waxahachie Tap, the SP runs nearly straight as an arrow through gently-rolling farm land, much of which is devoted to cotton. Five GE's and a lone EMD pilot SATX hoppers toward Fort Worth. Near Boyce the 5942-5578-5333-7211-5309-5310 pass a deserted farm house on a hazy February 1983 forenoon.

Upper left . . . On December 22, 1981, five green BN's work west toward Fort Worth with the coal empties.

Lower left and on this page . . . The same SATX train clumps through Waxahachie on SP rails, at the Fort Worth & Denver crossing. GE's 5916, 5937, 5536, 5073 and 5041 have up to 15,000 rated horsepower available. SSW caboose 47 trails the hoppers.

Upper left . . . Burlington Northern pusher 7847 shoves against an SSW caboose and the SATX loads, two miles east of Midlothian on a January 1983 afternoon.

Lower left . . . A westbound cab hop zips past long-shuttered Tower 94 and across the Santa Fe in Midlothian, Texas. Dark skies and occasional drizzle add to the mood of the abandoned tower as GP9E 3845 and BN SD40-2 7083 clatter by. SP 3845 started its career in 1959 as Texas & New Orleans 452.

Below . . . A quartet of Burlington Northern SD40-2's, and a lone GE, drift west and downgrade out of Midlothian on a hot July morning. For a loaded coal train going in the opposite direction, this is one of the toughest grades on the Fort Worth Branch. The climb to Midlothian is a prime reason for a helper; keeping the slack from running out (and pulling a drawbar), as the head-end crests the grade (near the Santa Fe crossing) and starts downhill.

SOUTHERN PACIFIC
9255
SOUTHERN PACIFIC
9253

7163
7163
7163
SOUTHERN PACIFIC

Left . . . Twenty of the many SP and SSW tunnel motors built by EMD in 1972 were placed in railroad class EF636-9 and given road numbers 9241-9260. Two of them, 9255 and 9253, are joined by BN 5009, as the trio handles SATX west approaching Mansfield. Fresh ballast is evident as upgrading continues on the Fort Worth Branch in 1982.

Below . . . Empty hoppers fade into the dust and haze as BN 7874-7816-7222-5383-7086 lope west near Mansfield with a mid-July version of the coal shuttle.

Lower left . . . At the same location, green BN 7163 is pushing east on the rear of a train of red-ended SATX loads. The tough grind up to Midlothian still lies ahead.

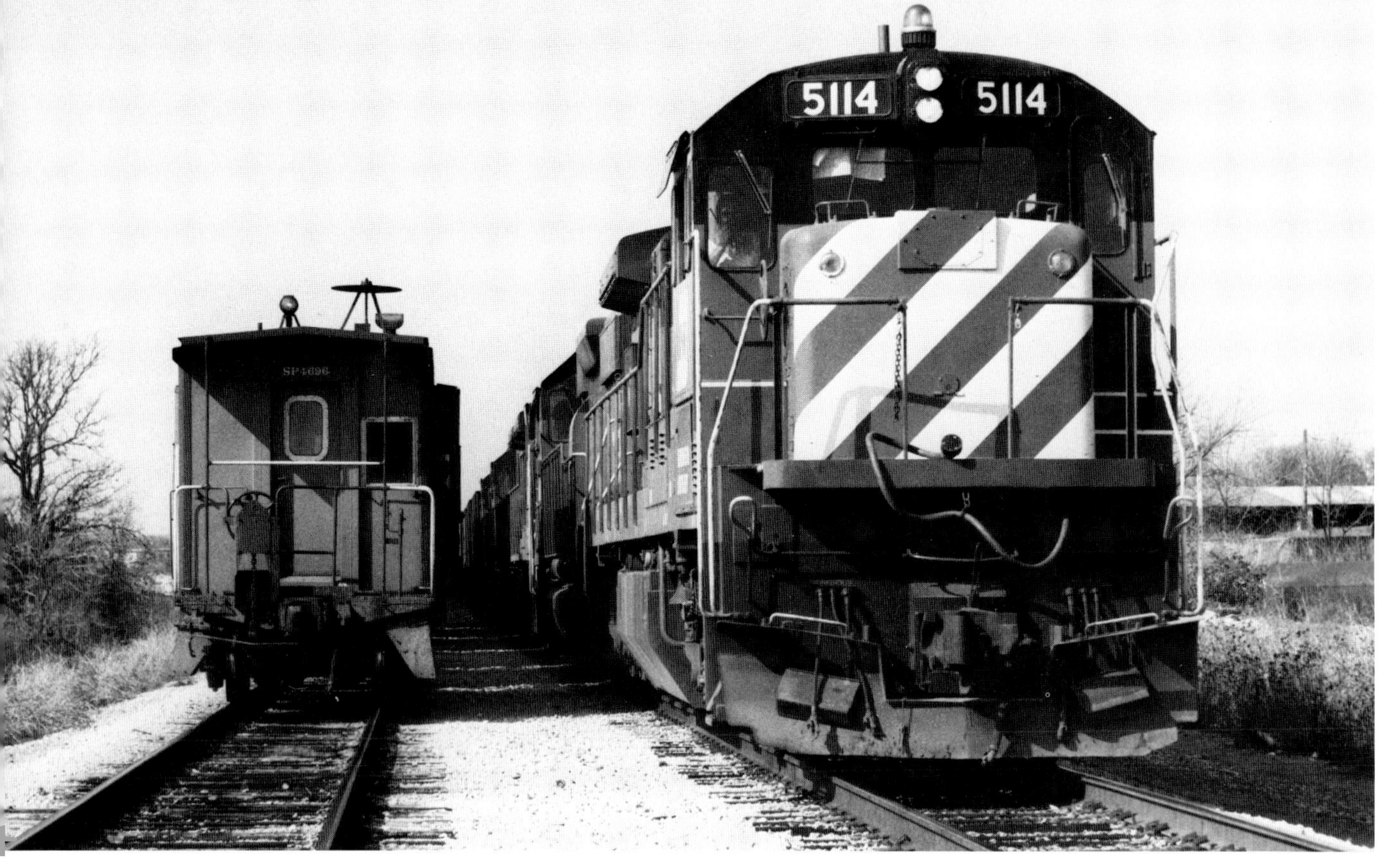
5114
5114

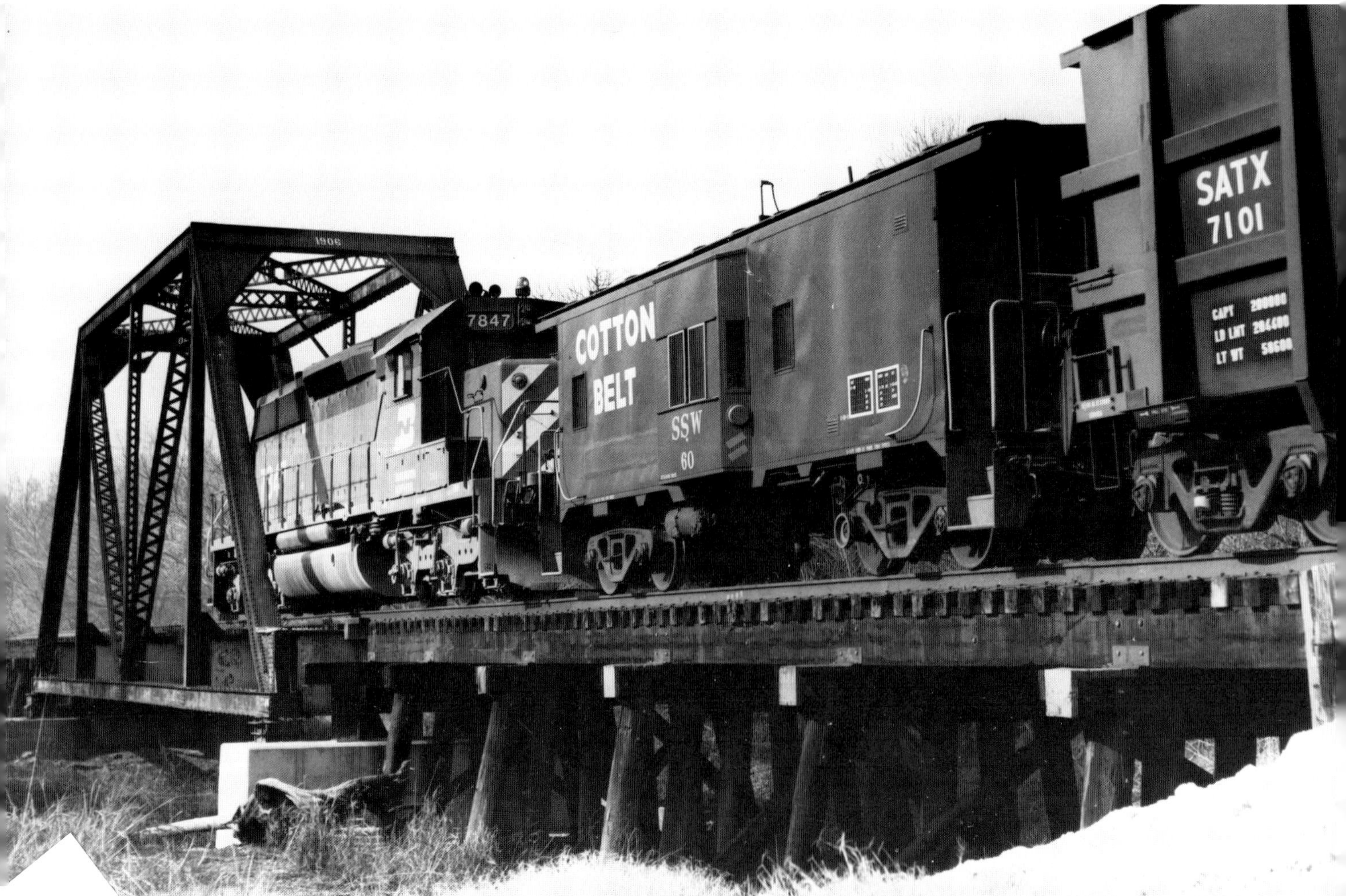
7847
COTTON
BELT
SSW
60
SATX
7101

Left . . . Now that the westbound Ennis to Fort Worth Turn has cleared on the main, BN 5114 and a January 1983 SATX are getting underway from the lengthy siding at Bisbee, Texas.

Above and lower left . . . The drum of steel on steel disturbs the tranquility of the cool January afternoon as the loaded SATX unit train enters Mansfield. Up front, GE 5114 leads EMD 7855 and three more GE's into town, while shoving hard on the rear is SD40-2 7847.

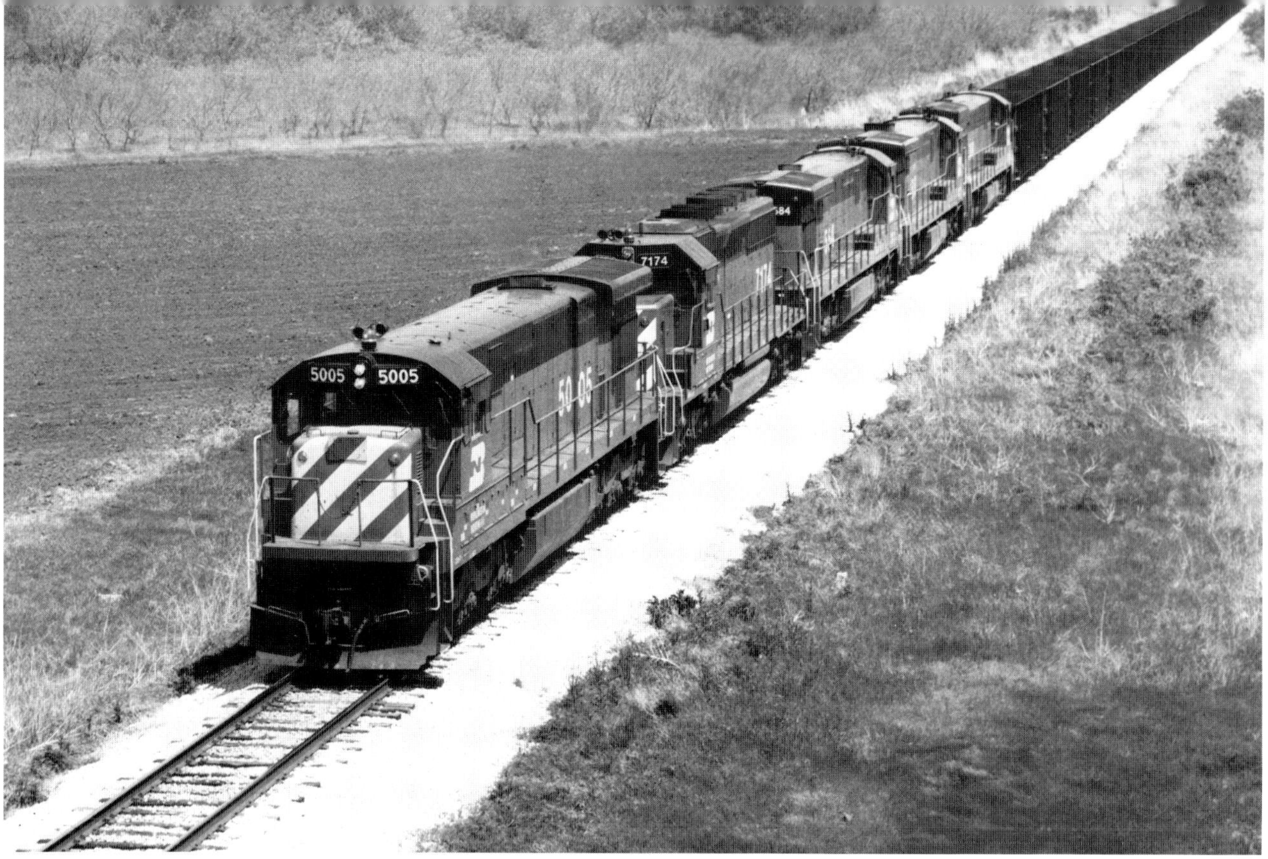
5005 5005
7174

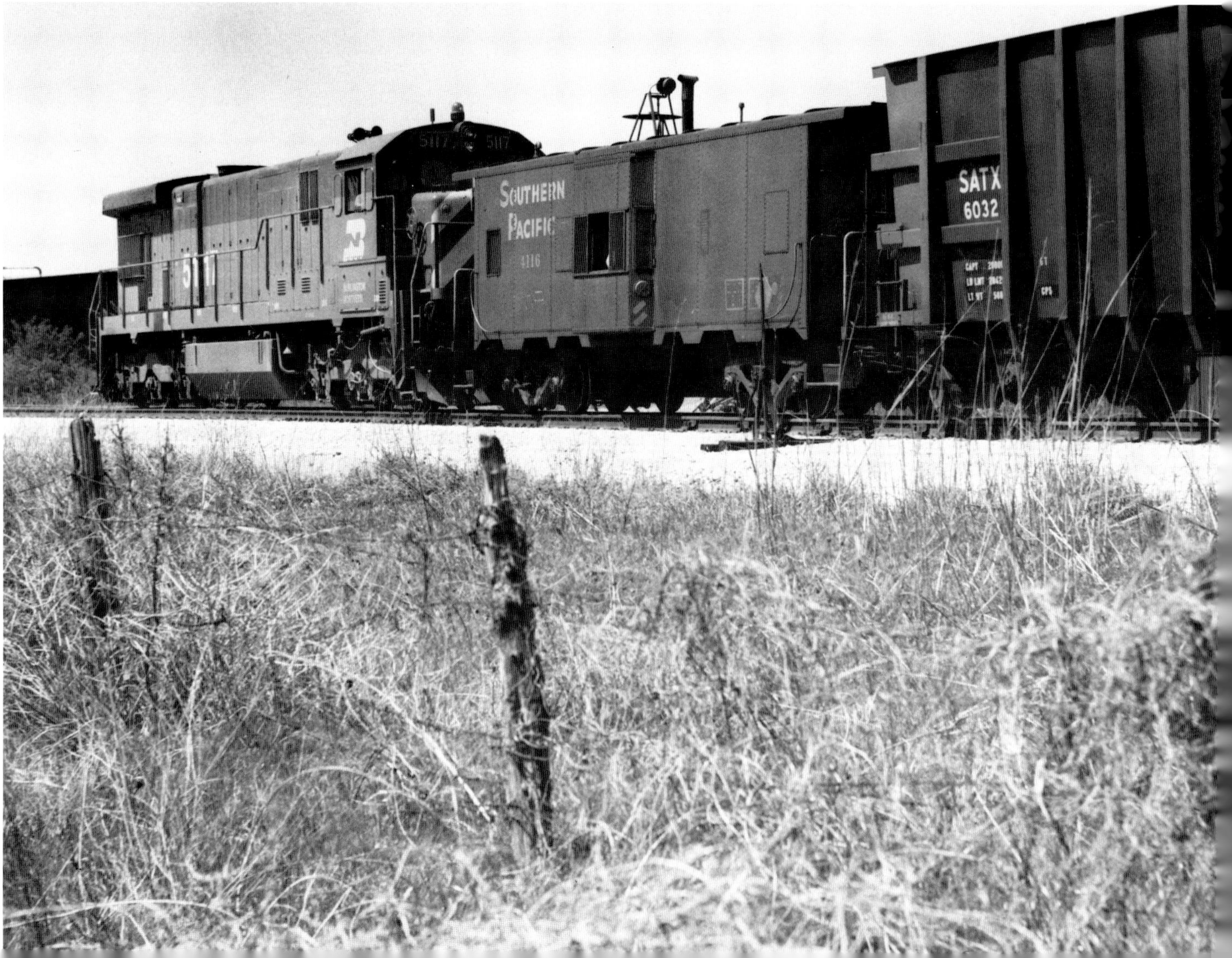
5117 5117
SOUTHERN PACIFIC
4116
SATX
6032

Left . . . Burlington Northern C30-7 5005 rolls SATX west approaching Mansfield, Texas, with the help of EMD 7174 and three more GE's.

Above . . . 5005 pulls the empties through the siding at Bisbee where they meet BN 7259, holding the main with a loaded train.

Below and lower left . . . After the empty hoppers disappear toward Fort Worth, 1980-built 5117 bunches the slack as the GE helper urges the loads toward Ennis, and eventually San Antone.

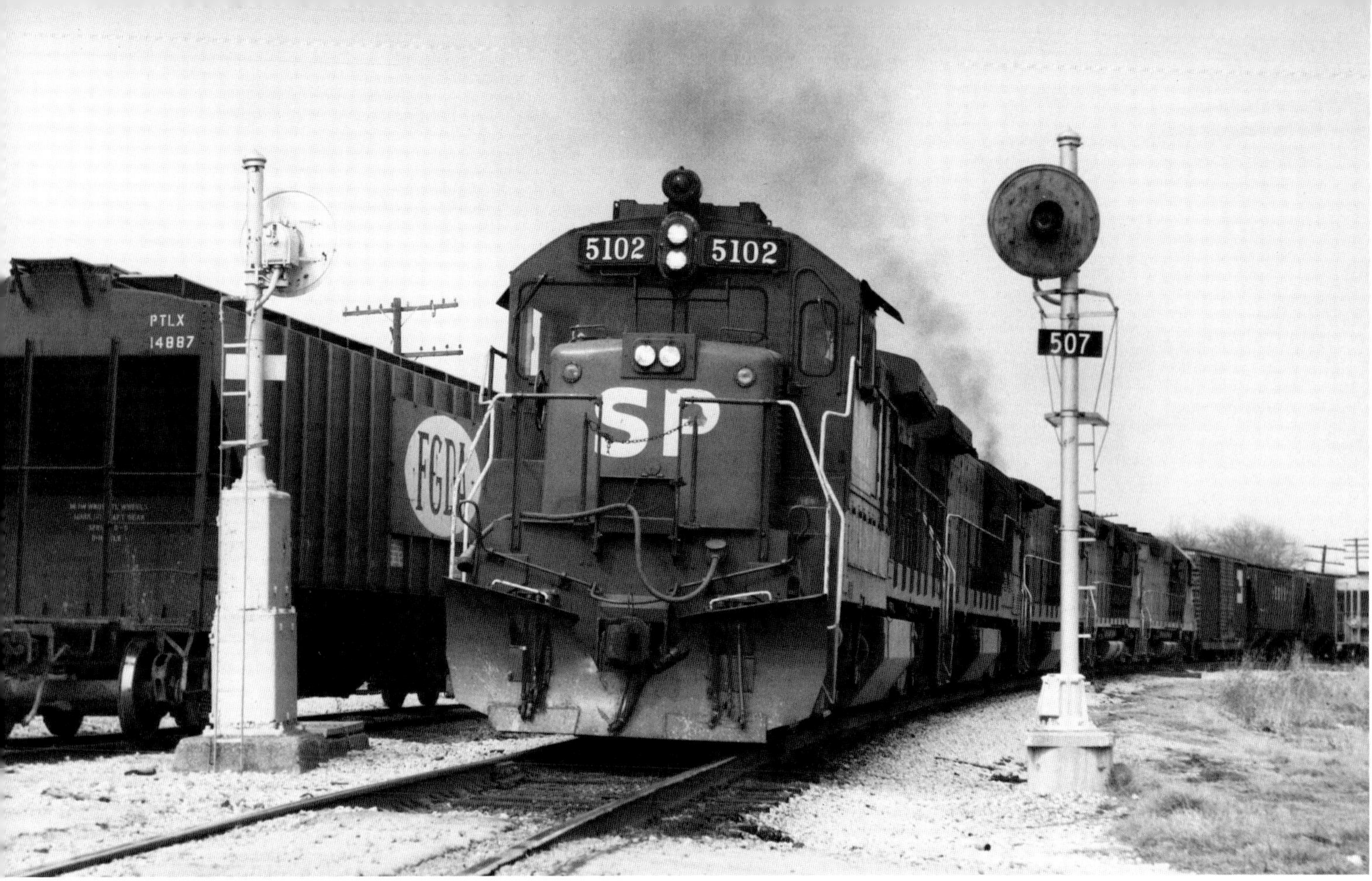

PTLX
14887
5102
5102
SP
507

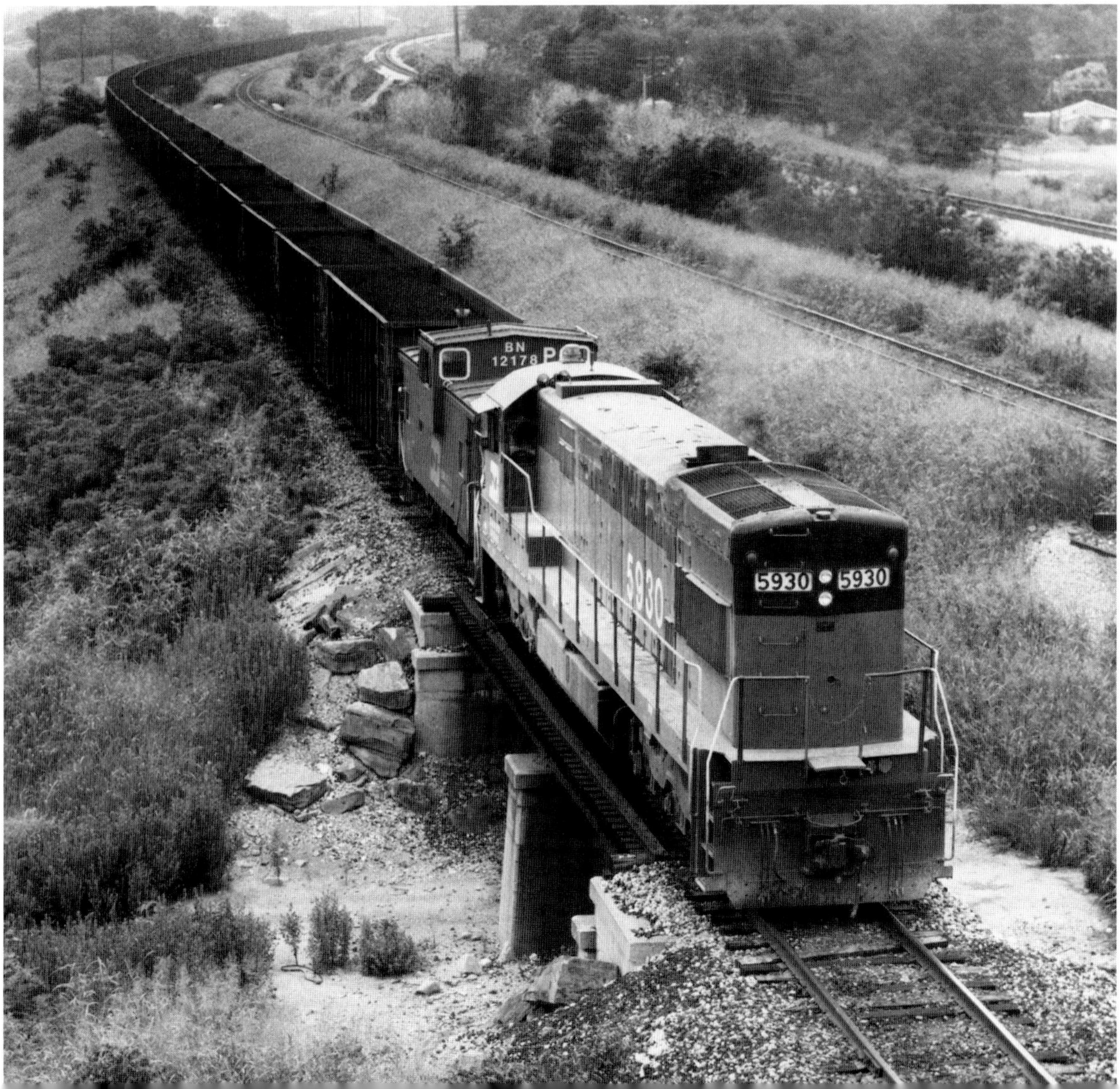

BN
12178
5930
5930
5930

Above . . . Cotton Belt 6508 is on the point of the power trio curving the return leg of a local Turn through residential Fort Worth. The nearly two-decade old GP35 (built in 1964) sports a white light, instead of the more normal red one, at the top of its triangular nose headlight arrangement. 6508 was originally SSW 768.

Lower left . . . A former Locotrol remote unit, GE 5930, has drawn the helper duty on a muggy June 1983 morning in Texas. The 1974-built BN U30C is shoving the SATX loads toward downtown Fort Worth where the coal train will move onto SP rails.

Upper left . . . At Jessamine Street in Fort Worth, three of the fifteen 1980-built B23-7's, 5102, 5113 and 5101, join with EMD's 6514 and 6658 on the Turn from Ennis. On this February Sunday afternoon, the five units still will struggle with the hill into Midlothian due to a longer-than-normal train.

7254
7254

Left . . . Valentines Day 1982 finds Burlington Northern's 7254 about to hammer the MKT diamond just south of Ney Yard in Fort Worth. SATX loads trail the 1980-built SD40-2.

Above . . . A few blocks further southeast, at Berry Street, the 7254 has been cut off and GE 5140 now is on the point. The SD40-2 will be tacked onto the rear of the consist as a helper engine all the way to Ennis.

Upper right . . . It is a corollary of Murphy's Law that when two single-track railroad lines meet (one of which has a light traffic density), a train will appear on both lines at the same time. May 8th in Fort Worth is no exception. SP's 5106, 5100, 5112 and 5109 are creeping up to a red board as three Santa Fe CF7's, and a second-generation EMD, wheel north with a green at Allen Avenue.

Above . . . After the red board clears, headlights flick on, and B23-7 5106 leads the local Turn across the AT&SF and into the nearby SP yard. Normally (in 1983) the Turn gets out of Ennis around 9 A.M., and departs Fort Worth between 1 P.M. and 3 P.M. Two different styles of lettering form the "SP" on the GE's short hood.

Lower right . . . At the same Allen Avenue location in Fort Worth, BN's Cascade green and white 5930 shoves SATX out of town. The Santa Fe main is in the foreground.

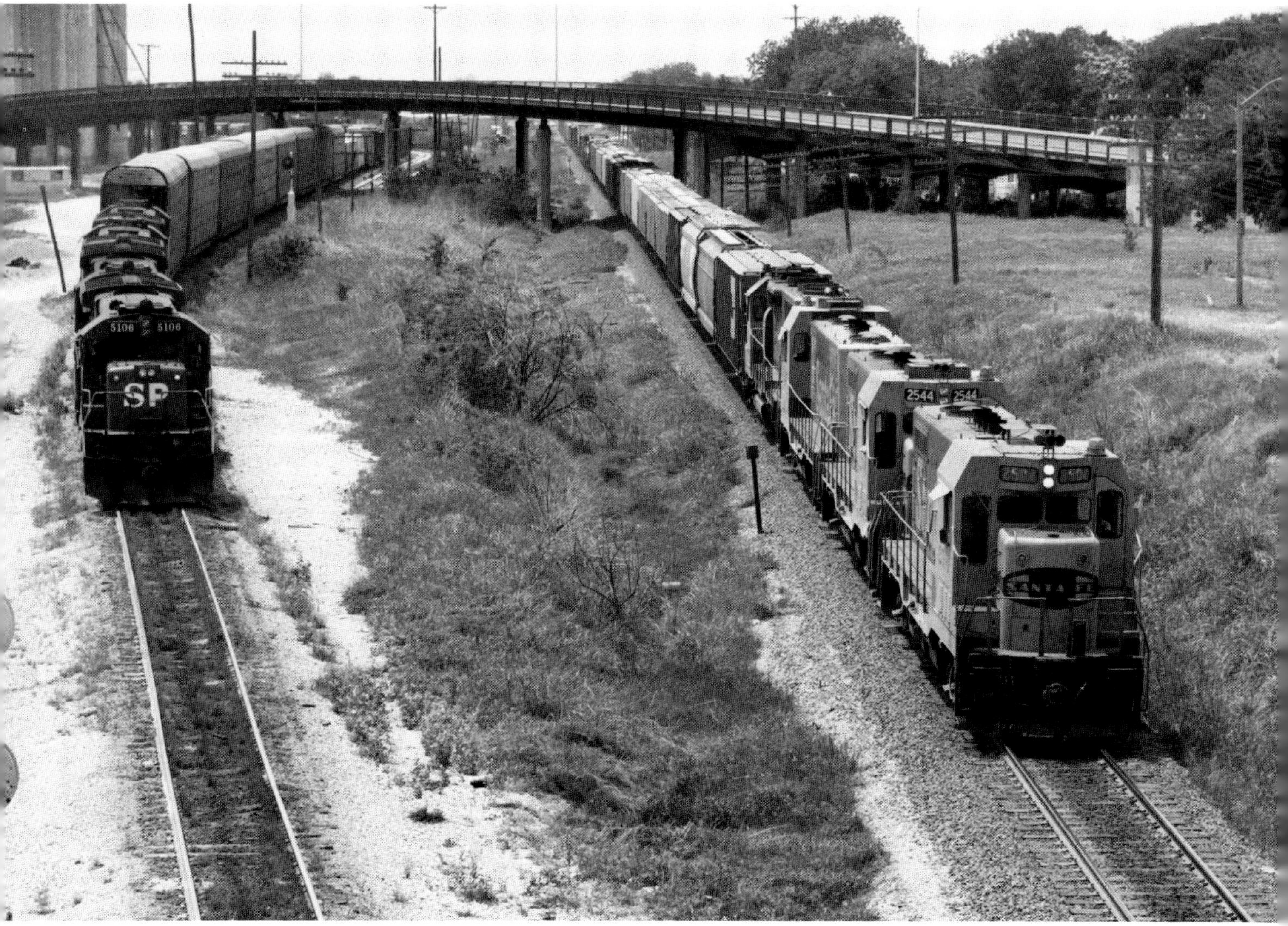

5106
SP
2544
SANTA FE
5930
BURLINGTON NORTHERN
BN
12178
BURLINGTON NORTHERN
SATX
3033

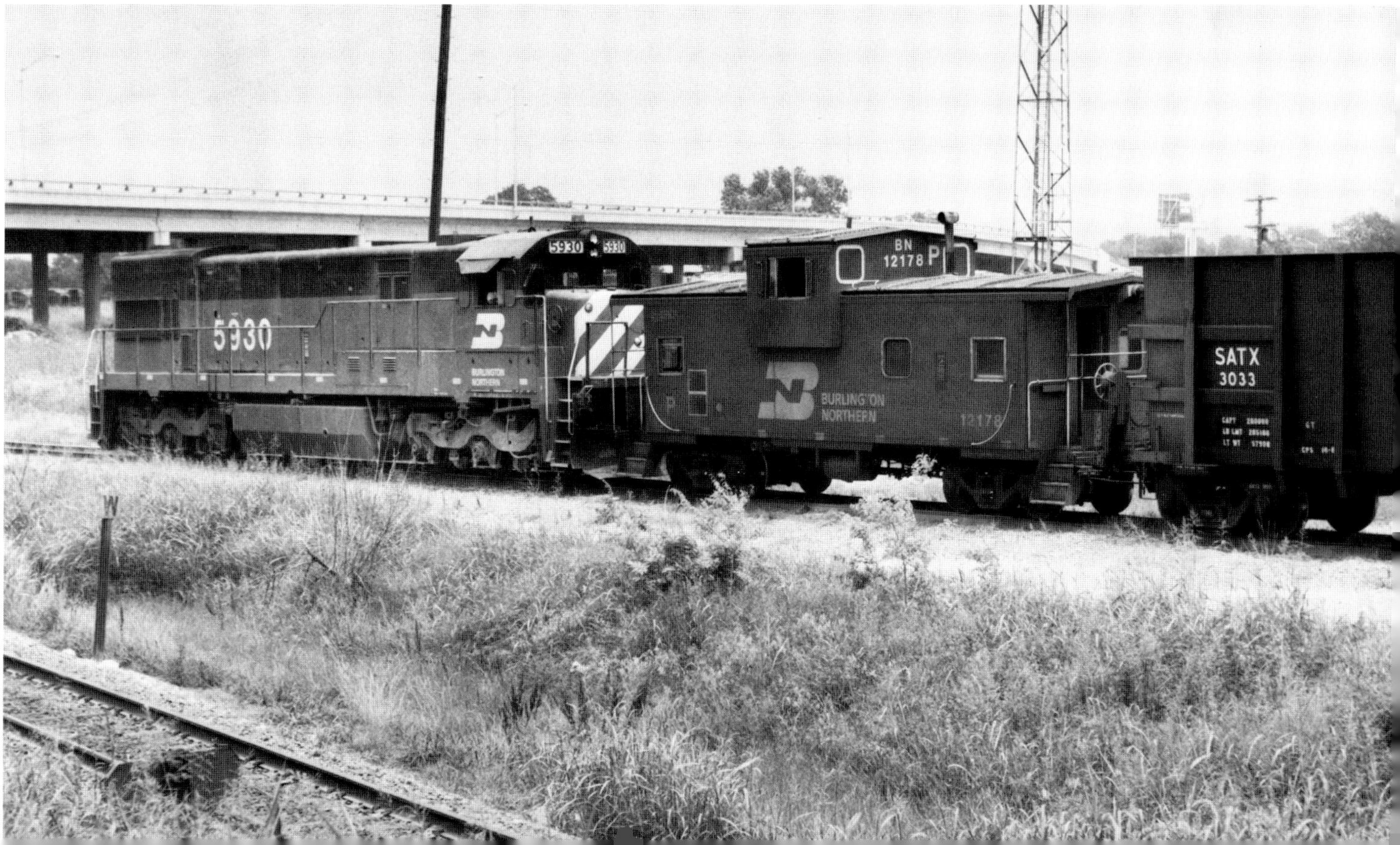

ENNIS-DALLAS

Upper left . . . MoPac 3222 and 3283, along with SP 4108, are working into Garrett at sunrise on the twenty-six mile haul from Miller Yard in Dallas to Ennis.

Lower left . . . As the trio pulls into the yard at Ennis, SP 7614 prepares to get underway for Fort Worth.

Right . . . The cowboy-hatted engineer is easing a big tunnel motor (SD40T-2 8524) into Ennis Yard in January 1982. Roots are strong in Texas, where Bob Wills and "Faded Love" and Patsy Cline's "Sweet Dreams of You" still mingle with the latest songs from the sons and daughters of the original C&W greats.

Below . . . Trailers bound for Dallas are urged out of Ennis by SD45T-2 9200 and SP's first freight-service GP40-2, the 7608. The morning sun gleaming off the flanks of the EMD's will soon turn this below-freezing December dawn into a more pleasant day.

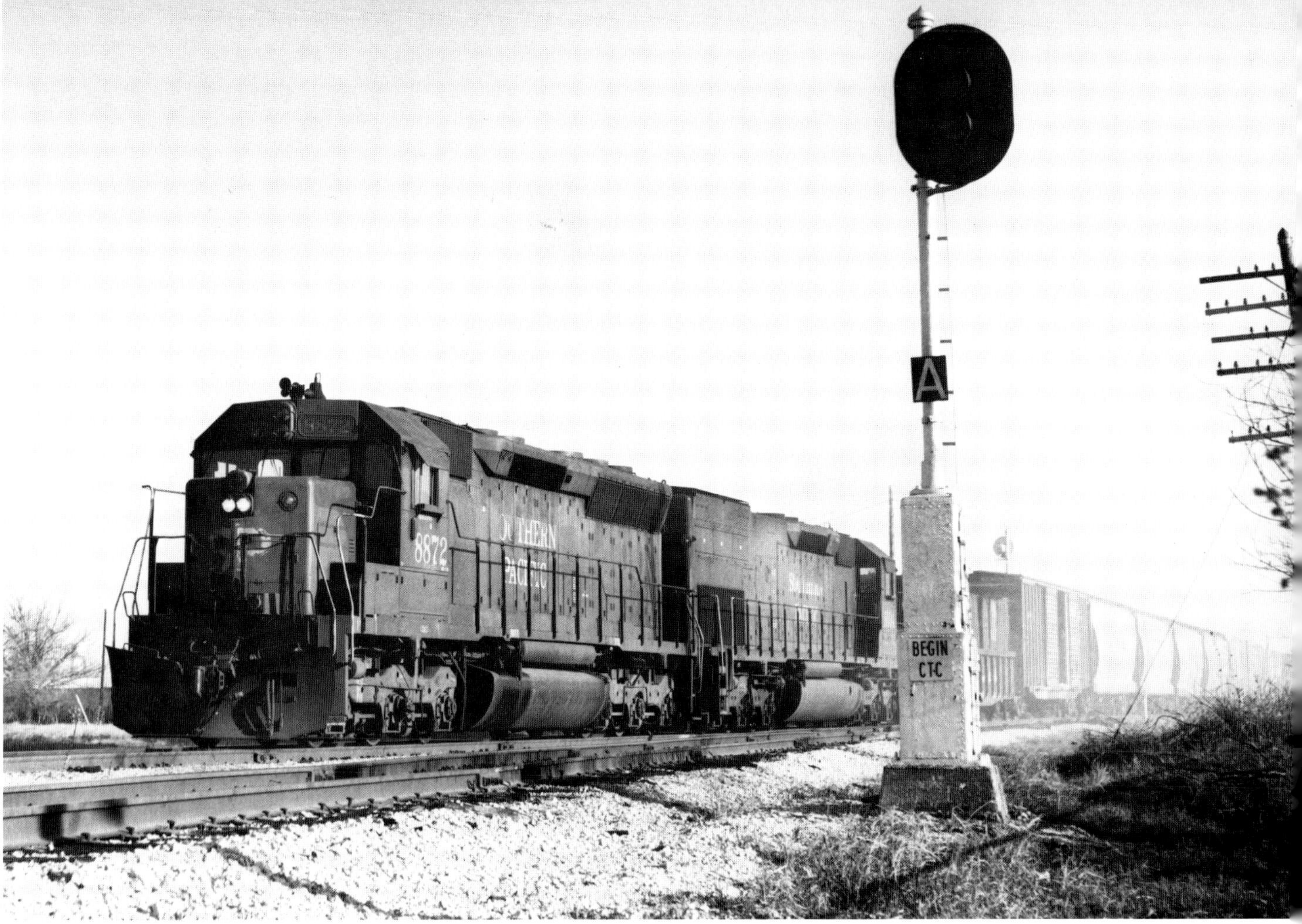
A
8872
BEGIN
CTC

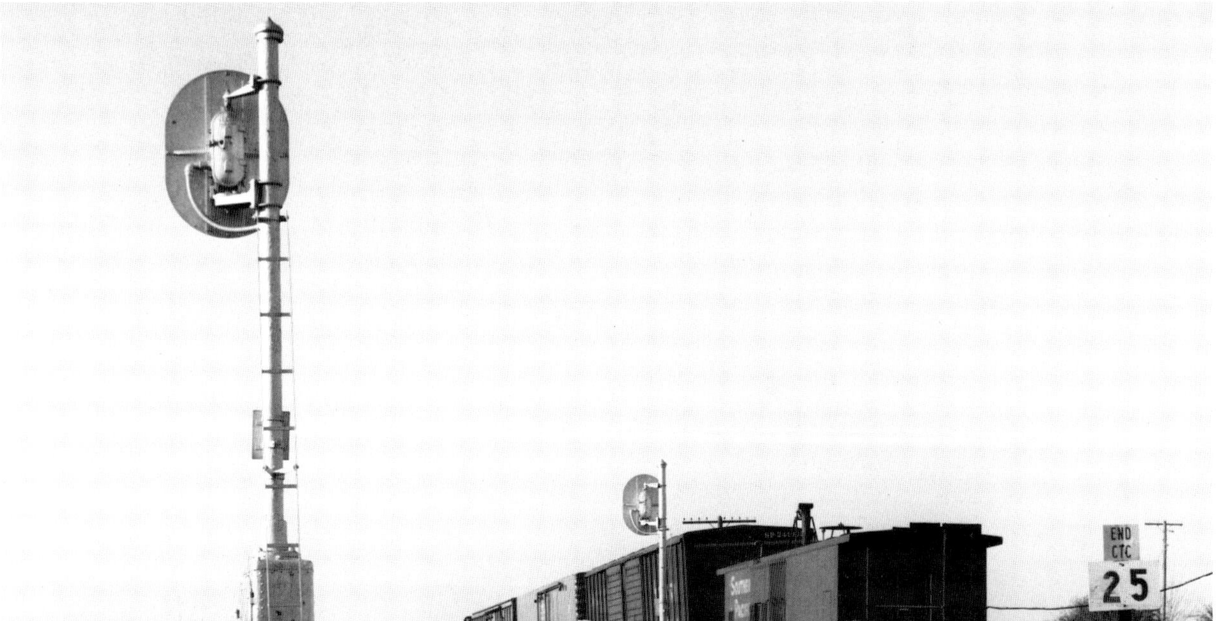
END
CTC
25

Above . . . Towering summer cumulus clouds are building as SD40T-2 8499 brings its Dallas-Ennis freight past the junction at Garrett.

Left . . . SD's 8872 and 9194 are raising the dust as they charge through Garrett, bound for Miller Yard in Dallas. Fort Worth Branch rails are in the foreground.

Above . . . GP9E 3794 is curving uphill into Garrett through an April rain in 1982. The GP9 came from EMD in 1956, assigned to the Texas & New Orleans as their 448, and was upgraded at the Sacramento Shop in 1976. Other units in the power team are GP9E 3777, B30-7 7799 and B23-7 5113.

Right . . . Another southbound from Dallas (eastward by timetable) is slogging up the short grade extending from south of Palmer to Garrett. SD40T-2 8499 and GE 5104 provide the muscle on a steamy June afternoon in 1982.

8499
SOUTHERN PACIFIC
SP

8499
SOUTHERN PACIFIC
SP

On December 19, 1982, one of the SP's road slugs was working an Ennis-Dallas turn. Teamed with Tractive Effort Booster Unit (TEBU) 1604 are GP40-2's 7948 and 7949. The northbound leg of the turn additionally drew B30-7 7829, seen *above* at Palmer. Returning in the afternoon, the slug trio curves into Palmer (*below*) and climbs upgrade toward Garrett (*upper right and lower right*). Twenty of the forty GP40-2's which came in 1980 were ordered with special equipment to operate in conjunction with road slugs, and numbered 7940-7959. Following slug prototype 1600, built by Morrison-Knudsen, SP's Sacramento Works produced 1601-1609 in 1981, utilizing the chassis from former U25B's as a base. U25B 6744, originally delivered to the SP in 1964, became the 1604.

Above . . . About two miles south of Ferris, Texas, a trio of 3600-horse EMD's, led by 9222, clatters across one of several wood trestles along this part of the Ennis Sub.

Upper left . . . Three more EMD's head north into Palmer in November 1981. SD40E 7354, leading the way, was originally 8410, before its Sacramento upgrading earlier in 1981.

Lower left . . . A rather grimy 9116 and GE's 7807, 7828 work uphill into Palmer in December 1982, enroute from Dallas to Ennis.

Above . . . Dallas-bound tunnel motor 9348 and B30-7's 7828, 7782 roll across a trestle north of Palmer in January 1983.

Lower right . . . 8524 is in the passing track at Ferris on a chilly January 1982 morning, as SSW 9375 approaches with *LADAT* (Los Angeles-Dallas Trailers).

Upper right . . . Underway again after the meet, 8524 and mates pass the former depot at Ferris.

8524
8524
SP

9375
SOUTHERN
PACIFIC

Upper left . . . Cotton Belt 9295 speeds toward Dallas as a January sun fades behind gathering clouds. Tomorrow will bring a sheet of ice to the area. The SSW SD45T-2 was built in 1973.

Lower left . . . 9348 splits the signals entering Ferris, Texas with a morning run to Dallas. Locomotives 9344-9370 were the last SD45T-2's lettered for the SP (9371-9404 are SSW units). All were built in 1975.

Above . . . A pair of 1972 tunnel motors, 9211 and 9176, both in class EF636-7, team up to move tonnage toward Dallas in December 1982. This particular trestle is a mile or so north of Ferris.

Above . . . A half-mile north of Wilmer, SD45 9118 goes it alone with trailers for Dallas in early 1982.

Upper right . . . 9211 and 9176 are north of Hutchins, Texas, as they journey toward Miller Yard in Dallas.

Lower right . . . SD40E 7308 has paused in its journey to Ennis to work a lumber yard on the north outskirts of Wilmer. A warm Texas sun bathes the scene, even in December. SD40 8461 emerged from the Sacramento program as the 7308 in July 1980.

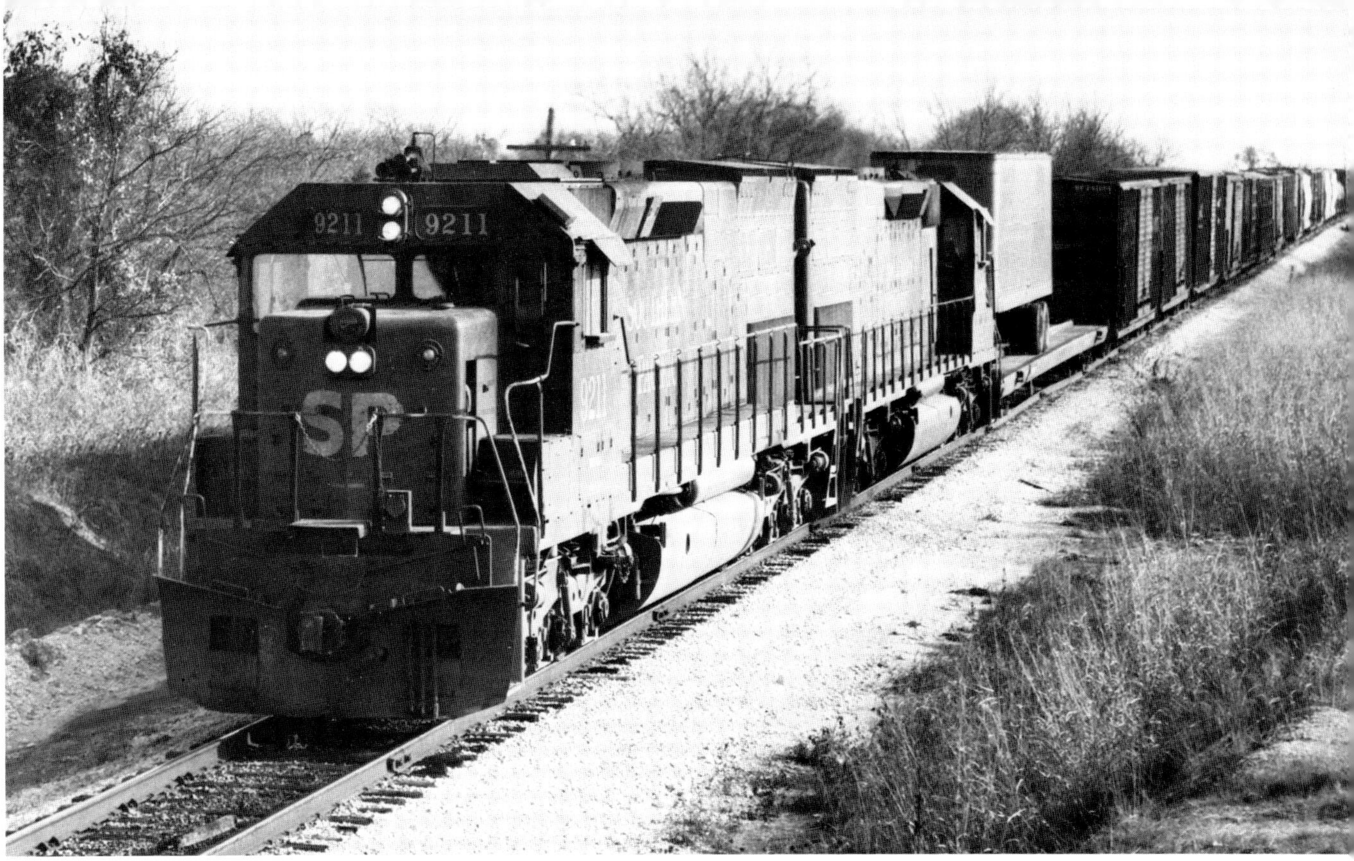
9211 9211
SP

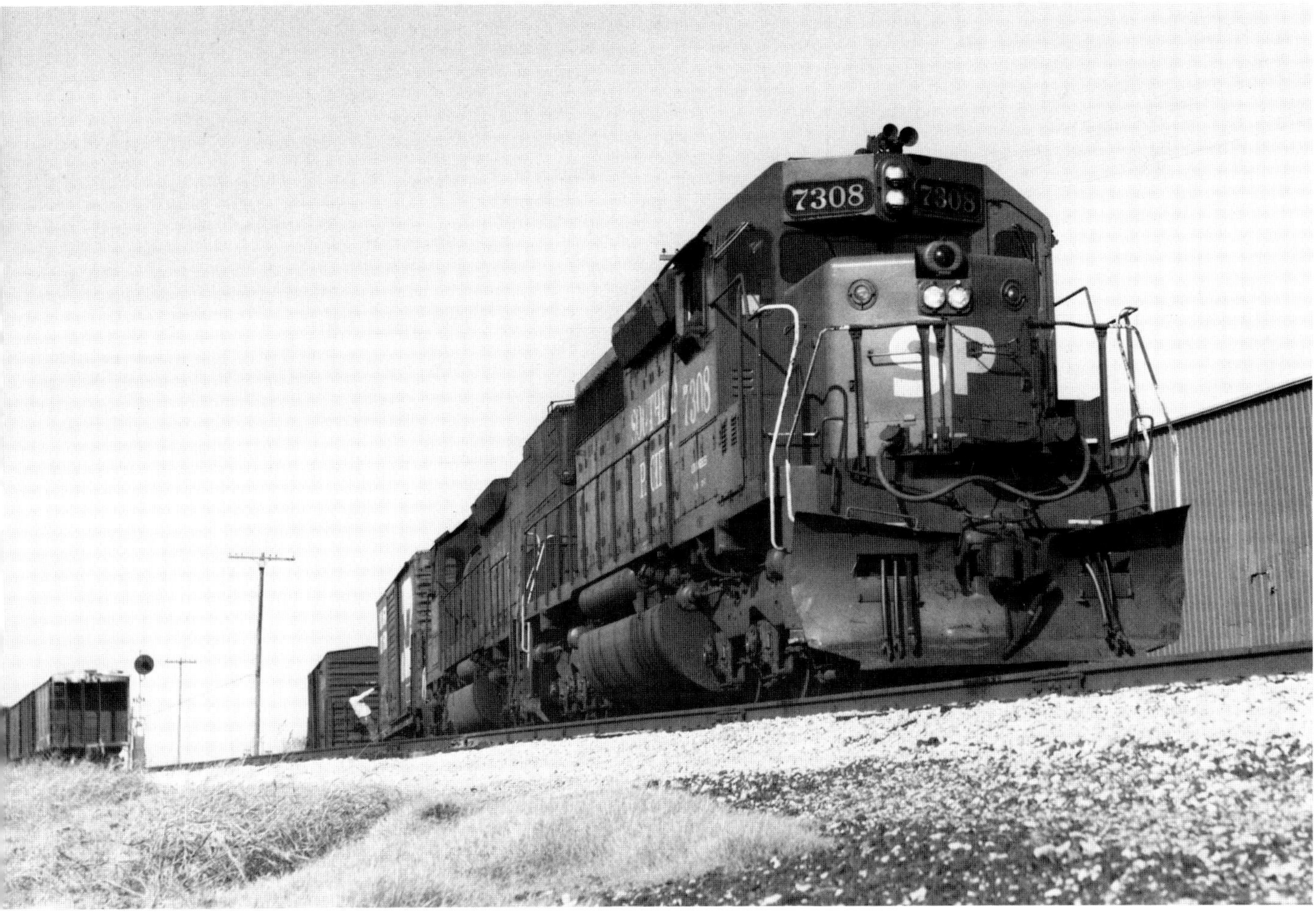
7308 7308
SP
7308

SP tunnel motors 9211 and 9176 approach Miller Yard in December 1982. There's been a lot of change in Texas railroading in the three decades since Hank Williams wrote about "that lonesome whistle blowing on the midnight train". Tunnel motors, and fleets of GP40-2's and B30-7's, now roll down the same rails through Ennis, Corsicana and Hearne once burnished by streamlined steam and PA's on the *Sunbeam* and the *Hustler*. The Southern Pacific's increasing use of the old Rock Island "Golden State Route" across Kansas has affected traffic on the Texas lines. But if business perks up through the mid and late 1980's, there will continue to be plenty of interesting SP railroading in the Lone Star State.

SP SD45 9116 leads a five-unit power team in command of a December 1982 *BSM* curving downhill beyond Angus. Houston & Texas Central rails were spiked down northward through this region 111 years before the big red and gray EMD raced south.